ESSEX ROCK

GEOLOGY BENEATH THE LANDSCAPE

Ian Mercer & Ros Mercer

Illustrations and design by Trevor Johnson

Pelagic Publishing

Published in 2022 by
Pelagic Publishing
20–22 Wenlock Road,
London N1 7GU, UK

www.pelagicpublishing.com

Essex Rock: Geology Beneath the Landscape

British Library Cataloguing in Publication Data
A catalogue record for this book is available from the British Library

https://doi.org/10.53061/YKSO7005

ISBN 978-1-78427-279-1 *Paperback*
ISBN 978-1-78427-280-7 *ePub*
ISBN 978-1-78427-281-4 *PDF*

Cover artwork by Peter David Scott. Summertime in Essex during the Anglian glaciation 450,000 years ago. Steppe mammoth *Mammuthus trogontherii* is four metres tall at the shoulder and weighs ten tonnes. The ancient Thames flows across mid-Essex, where Chelmsford is today.

Contents

Preface to the second edition

The original edition of this book, written by Gerald Lucy, was the first general account of Essex geology for a wide readership. In the years following that pioneer edition, the deep-time story of Essex has been further enriched – especially its ice age history – as we ourselves discovered following our retirement. Much of the 'rock' of Essex Rock is admittedly rather squashy and crunchy, but this does not deter geologists from defining it as 'rock'. When we were both at university in the 1970s, Essex geology was dismissed as 'gardening'. Like Rev. Eley, vicar of Broomfield, who in 1859 wrote Geology in the Garden, we looked more closely at the pebbles all around us. There is a story in every single one; we wanted to learn more and to share these stories.

All of us are living in an ice age – something that has profoundly affected Essex over the past 3 million years. Yet this county was born more than 500 million years ago. Our deep history begins when the land we know as Essex was part of a large continent close to the South Pole. Looking around you anywhere in Essex, along footpaths, beaches, urban areas or village streets, we want to help you notice some of the details and features in the landscape and link them to this deep-time story. You might see a strange boulder on a village green, or layers of sand or chalk in a new road cutting. There are interesting pebbles in church walls, in gardens and fields – as well as along the seashore. These can all bring geology to light in unexpected and fascinating ways.

Research by professional geologists and archaeologists and many amateur fossil collectors and other enthusiasts continues to increase our understanding of time and of environmental change. Building and engineering projects such as tunnels, coastal defences and structural foundations require continuing investigations into the nature of the rocks and their geological stories. All of this is adding constantly to the fascinating history of Essex , enriching our knowledge and opening up new avenues to explore.

We hope this new edition of Essex Rock will encourage more people to enjoy the land and its story of time – past and future. Such enjoyment is enhanced through individual enquiry and the use of museums and other institutions as well as libraries and clubs. Amid rapid changes, now is the time to appeal for more examples of the 'real thing' in museums and to press for a dynamic treatment of geology in museums and schools. Many museums have lost their displays of useful geological detail and accessible reference material and now have few or no staff who can identify geologically related objects and materials. Those displays that remain need careful protection. The national curriculum for schools in England currently contains little coherent geological content. We hope this book will help promote education and enlightenment about rocks, land use and soil, water supply, oceans and climate; that it might help people in some small way to comprehend the huge changes to come.

This account of the county might also encourage you to promote the conservation of important geological sites – currently low in the list of priorities in planning and development – you can find out more by joining clubs and societies such as the Essex Field Club, Essex Rock & Mineral Society, or a local U3A group and by viewing the work of the conservation group GeoEssex online.

Ian Mercer and Ros Mercer
2022

Preface to the first edition

Many people have visited Walton-on-the-Naze and have found sharks' teeth and fossil shells on the beach beneath the Naze cliffs. A number of these visitors must have wondered how they came to be there, how old they were and what Essex was like when these creatures were alive. Until now this information has been very difficult to obtain; no popular works have ever been written about Essex geology, and the subject receives little attention in natural history publications. This book, originally produced by the Essex Rock and Mineral Society, is intended to fill this gap and provide an account of the prehistory of the county up to the time of man.

Few people think of rock when they think of Essex, yet every landscape is built on rock of one or more kinds, from granite to the softest clay or sand. Each piece of rock is a store of prehistory. Even a pebble from the garden has its own story to tell. We therefore hope that this book will generate more interest in geology, the subject that investigates the foundations of our natural heritage. Although Essex has few prominent geological features when compared with most other parts of Britain, there is still much to see and this book emphasizes the visible evidence of the county's geological past. Geology has been called 'the great detective science' and it is always best to examine the evidence for yourself first hand.

What started off many years ago as an idea for a slim booklet on Essex fossils has grown into the present work out of a desire to produce a book which would not only be of interest to the general reader, but also be a useful source of reference for anyone interested in natural history. So as to present this subject to as wide a readership as possible, some generalisations have inevitably had to be made and some geological terms that may now be considered outdated have been used. Comment or criticism will be welcome and considered for incorporating in any future edition. The text, diagrams and illustrations have been compiled from a large number of sources, and jargon has been kept to a minimum.

For the purposes of this book Essex is taken to be bounded on the north by the River Stour, on the south by the River Thames, on the east by the North Sea, and on the west by the River Lea. It will be clear from this that we have chosen the old geographical borders of Essex before the local government reorganisation of the 1960s, and so include the present London boroughs of Barking, Havering, Newham, Redbridge and Waltham Forest.

Gerald Lucy
1999

A look beneath the Essex landscape

Foreword

As the 841st High Sheriff of Essex, I am delighted to be supporting this excellent and much-expanded new edition of *Essex Rock*. Its chapters cover in detail how Essex has changed over past ages, and this is backed up by many photographs, illustrations and maps.

As an Essex farmer, I have found fossil echinoids on the surface of the soil as well as worked flints dating back to the Hoxnian interglacial. Five to six metres below the topsoil I have uncovered the remains of woolly mammoth dating back 40,000 years to the most recent glacial stage, the Devensian.

Anyone who has an interest in Essex and its landscape and the geology beneath this landscape should read this book. It gives us an insight into how complicated and varied the geology of Essex is, and how climate has affected this landscape over millions of years. From rocks to dinosaurs, from seas to chalk, giant sharks to mammoths, palm trees to the ice age, it is all in here and much more.

As new research papers have become available from experts across many fields, these have been used to update this edition. Furthermore, all manner of professional geologists, scientists, archaeologists, and many amateur fossil collectors and other enthusiasts have added to this book too.

As the awareness of climate change is greater than ever before, it is important that we understand the past to help us go forward in the future. *Essex Rock* helps answer some of these questions and might make the reader look differently at the world around them. Thank you to everyone who has added to this story in *Essex Rock*.

Simon Brice
High Sheriff of Essex 2021–22

ESSEX ROCK
GEOLOGY BENEATH THE LANDSCAPE

PELAGIC
PUBLISHING

IAN MERCER & ROS MERCER

Acknowledgements

We are indebted to all those who helped with the writing and production of the first edition of this book in 1999 and who are given acknowledgement in that edition. The first edition not only received wide acclaim but also won an award in the Essex Book Awards; credit for this is due to everyone who was involved, particularly Clive Walpole for the illustrations. This new edition has similarly received support from a large number of people and has taken advantage of new technology, such as the use of colour photographs and illustrations, which wasn't possible for the first edition.

Many of the diagrams are based closely on the work of geologists who are at the forefront of modern and detailed research, notably that of Prof. David Bridgland of Durham University, who had, for instance, depicted the 'staircase' sequence of gravel terraces of the Thames and Medway Rivers, upon which our block diagram sequence in the ice age chapter is based, among many other instances. The careful guidance of Dr Peter Allen is also most gratefully acknowledged. Without the guidance of such dedicated scientists, the accuracy and veracity of text, reconstructions and diagrams would have suffered greatly. Nevertheless, we fully accept that we may well have included our own mistakes and misapprehensions and we take full responsibility for any such errors.

We are glad that we were able to benefit from the pictorial insights and graphics skills of a talented illustrator, Essex Rock & Mineral Society member Trevor Johnson, who provided outstanding transformations of our diagrams and created the book design. The equally delightful dialogue with our cover artist Peter David Scott has been a revelation in how to enjoy geological depths of time brought to life realistically and vividly with a powerful artistic impact. Right from our first contact we received great encouragement from our publisher, Dr Nigel Massen, and Production Editor David Hawkins, who greatly eased the path through production.

Without enormous help from several donors, we could not have achieved what we had set out to do. We are particularly grateful to the Essex Field Club for their very substantial support with publishing costs in order to make our work available to the widest audience. Essex Rock and Mineral Society member David Grayston made it possible to complete all of our illustrations with a most generous donation. Further very welcome grants were put towards the cost of illustrations by Essex Rock and Mineral Society, The Curry Fund of the Geologists' Association, GeoEssex, the Quaternary Research Association, Leez Construction Company, Marden Homes, IBIS, G&B Finch and David Turner.

We wish to thank all those who have assisted us with discussion and in checking our work, and who have answered our numerous questions to enable us to compile maps, sections, tables and many of the other diagrams and artworks in this book. Any remaining errors or misinterpretations are entirely those of the authors.

We are grateful to at least the following very helpful people and no doubt many others who, we are sad to say, we may have failed to list throughout several years of dialogue: our greatest thanks go to all of you, even if you are missing from our list.

Dr Ken J. Adams
Dr Peter Allen
Dr Nick Ashton
Dr Haydon Bailey
Prof. David Bridgland
Simon Brice
Dr Malcolm Butler
Kathy Coombes
Ashley Cooper
Michael Daniels
Dr James Dilley
Richard Ellison

Dr Peter Frykman
Bill George
David Grayston
Tim Holt-Wilson
Dr Richard Hubbard
Jon Hughes
Jim Jenkins
Dennis Kell & Dr Anne Kell
Gerald Lucy
Bob & Caroline Markham
Prof. Rory Mortimore
Dr Tim Newman

John Ratford
Prof. Jim Rose
Jeff Saward
Prof. Danielle Schreve
Jonathan Spencer
Prof. Chris Stringer
Simon Taylor
David Turner
Susanna van Rose
Russell Wheeler
Prof. Mark White
Mark Woods

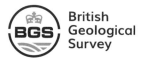

Geological maps

Geological maps have been assembled using information from various sources, notably from parts of the detailed geological map of the vice-counties 18 and 19 of Essex by Dr Ken J. Adams from *Flora of Essex* by Stanley T. Jermyn, 1974; also from areas within the BGS Geology of Britain online 3D viewer, together with detailed information from Prof. David Bridgland and a number of other experts and literature sources, particularly in the fields of Pliocene and Pleistocene deposits and their interpretation, such as the Quaternary Research Association's Field Guides to the area published in 2014 and 2019.

Essex Field Club

The Essex Field Club is a society for people with an interest in the natural history and geology of Essex and its study, recording and conservation. The Club was founded in 1880 'to promote the Study of the Natural History, Geology and Pre-historic Archaeology of the County of Essex and its borderlands; to establish a Museum and to issue publications'. Charles Darwin and Alfred Russel Wallace were amongst the founder members. The Club houses a large and important collection of Essex geological specimens, literature and images, of both historic and current importance, and holds regular field meetings throughout the county.

GeoEssex

GeoEssex is a geoconservation steering group, a partnership of local geological groups and other bodies with an interest in the geodiversity of Essex. It aims to promote, conserve and enhance geodiversity, involving local planning processes, education and public outreach.

Essex Rock and Mineral Society

The Society, formed in 1967 by Rowley Collier, is a club open to all who wish to enjoy and promote interest in rocks, minerals, gems, fossils and general geology through meetings, social media, an annual Show, displays, field-visits and other means. The Society published the first edition of *Essex Rock* in 1999.

We received much help from experts and derived many illustrations from those published in specialist publications and academic papers. Notable examples are as follows.

The image of *Carcharocles megalodon* in Chapter 8 appeared in the journal *Palaeogeography, Palaeoclimatology, Palaeoecology* 469, 1 March 2017: 84–91, 'Did the giant extinct shark *Carcharocles megalodon* target small prey? Bite marks on marine mammal remains from the late Miocene of Peru.' Authors: Alberto Collareta et al. Artwork by Alberto Gennari 2015.

The diagram for global climate changes in Chapter 13 is based on the CENOGRID diagram, climatic variability of the Cenozoic, compiled by Dr Thomas Westerhold, Director of MARUM, the research centre of the University of Bremen, with kind permission.

Aldiss, D.T. 2014. *The stratigraphical framework for the Palaeogene successions of the London Basin, UK*, Nottingham, British Geological Survey, OR/14/008.

Bridgland, D.R. et al. 1994. *Quaternary of the Thames*. CGR Series No.7, Chapman & Hall.

Bridgland, D.R. et al. (eds) Quaternary Research Association Field Guides 2014 & 2019.

Cooper, Ashley. 1982. *The Long Furrow*; 1994, 1998. Heart of Our History vols 1 & 2, Bulmer Historical Society.

Ellison, R.A., Woods, M.A., Allen, D.J., Forster, A., Pharoah, T.C. and King, C. 2004. Geology of London. *Memoir of the British Geological Survey, sheets 256, 257, 270 and 271.*

Gale, A. 2020. Richard Granville Bromley (1939–2018) – Understanding the chalk. *Proceedings of the Geologists' Association* 131: 618–628.

King, C. 2016. A revised correlation of the Tertiary rocks in the British Isles and adjacent areas of NW Europe. Geological Society Special Report, No. 27.

Lee, J.R., Woods, M.A. and Moorlock, B.S.P. (eds) 2015. *British Regional Geology: East Anglia* (5th edn), Keyworth, Nottingham, British Geological Survey.

McKerrow, W.S. (ed.) 1978. *The Ecology of Fossils: An Illustrated Guide*, Duckworth.

Minter, Peter. 2014. *The Brickmaker's Tale*, Bulmer Brick & Tile Company.

Montgomery, D.R. 2017. *Growing a Revolution*, W.W Norton.

Mortimore, R. 2011. A chalk revolution: what have we done to the Chalk of England? *Proceedings of the Geologists' Association* 122: 233–297.

Musson, R.M.W. 1994. *A Catalogue of British Earthquakes*, British Geological Survey Global Seismology Report, WL/94/04.

Savrda, C.E. 2007. Ch. 9, pp. 149–158 in William Miller (ed.) *Trace Fossils: Concepts, Problems, Prospects*, Elsevier.

Sumbler, M.G. 1996. *British Regional Geology: London and the Thames Valley* (4th edn), London: HMSO for the British Geological Survey.

Woodcock, N.H. and Strachan R.A. 2012. *Geological History of Britain and Ireland* (2nd edn), Wiley-Blackwell.

Woods, M.A. and Lee, J.R. (guest eds) 2018. The Geology of England, *Proceedings of the Geologists' Association Special Issue* 129/3.

1

A picture of Essex countryside. Everything you see here is influenced by the geology beneath – with its story of deep time. This is a view in mid-Essex across the London Clay fields to Buttsbury church and beyond to the sand and gravel-topped ridge between Stock and Billericay.

The Age of the Earth

4,560 million years 4,000   3,000   2,000   1,000   541   Now

Origin of the Earth   Precambrian   ▲ The age of Essex ▲

Reconstructing Essex

Geology: the great detective science

If you walk through the Essex countryside it is hard to believe that in the distant past it was covered by a vast ice sheet or that mammoths once roamed this land. Moreover, at an even earlier time, the whole of what is now Essex was submerged beneath a subtropical sea in which sharks and crocodiles swam.

The study of these ancient times involves geology, the science concerned with the history of planet Earth and the study of natural processes that have affected it and, indeed, are still affecting it today. This book aims to reveal the unfamiliar and often surprising deep-time story of Essex.

Geologists learn about the history of the Earth by studying its rocks. It is not only the professional geologist who can make a contribution to the science. Many amateurs in Essex have discovered new sites for research, or found fossils that have turned out to be species new to science. For more than 200 years, amateur and professional geologists and other scientists have enquired into the land, collecting fossils and artefacts, and recording the rock layers. In the nineteenth century the fresh cuttings for the new railways revealed hitherto unseen geology. Until recent times there were many small pits to inspect

▲ Gravels at Bull's Lodge Quarry near Boreham laid down by the River Thames when it flowed through central Essex over half a million years ago.

▲ These fossil shells can be seen in the cliffs at Walton-on-the-Naze.

◄ Rock layers and the fossils within them help us to work out the story of this planet's past. This rock is known as Red Crag and was the bed of a shallow sea over 2 million years ago.

– those dug for brickmaking and lime burning, for gravel and sand for roads and walls. Through all these years, a developing story has been pieced together. To experienced eyes, rocks reveal a wealth of information.

During a visit to a quarry or a cliff, you might see rocks full of fossils, the remains and imprints of animals and plants that lived in the past. The study of the rock layers and the fossils within them gives information on the conditions that existed on our planet long ago, helping us to understand the history of life, to help predict climate change and to look after the planet more successfully.

The rocks beneath the Essex landscape have a fascinating story to tell. Knowledge of geology and its effect upon the landscape can add greatly to the enjoyment of the countryside, whether you are in Essex or any other part of the world.

Geological time

A feeling for the vast periods of geological time comes with investigating the Earth's deep past. Geologists talk of millions of years in the same way that archaeologists talk of hundreds or thousands of years. It is useful to compare geology and archaeology because the geologist is often mistaken for an archaeologist when they are found carefully studying a hole in the ground. An archaeologist studies the history of humankind through the excavation of evidence of human activity mostly from the last 500,000 years. A geologist, on the other hand, studies the history of the Earth since its formation 4,560 *million* years ago.

To put this vast span of time into perspective, if 4,560 million years is visualised as 24 hours, the last 500,000 years of human existence would represent less than 10 seconds. In attempting to grasp this huge time span it becomes clear that sufficient time has elapsed for enormous changes to have taken place on Earth. Geological time is divided into periods such as Jurassic and Cretaceous, just as English history is divided into periods such as Roman, Saxon and Norman. The geological story of Essex takes us back some 500 million years.

In an archaeological excavation, the deeper one digs the older the finds will be. Medieval coins, for example, will be found beneath Victorian bottles, and Roman pottery will be in a still deeper layer. This basic rule of archaeology parallels two important principles of geology: firstly that where there has been no disturbance, each layer is younger than the one that lies beneath it, and is older than the one that lies on top of it; and secondly, that it is usually possible to estimate a date for each layer from the relics it contains. The same principles are valid whether we are considering human relics buried in layers of soil or fossils preserved in layers of rock. Sometimes a rock layer is devoid of

◄ Gerald Lucy looking into Essex geology. Almost everything we know about the Earth's distant past comes from looking at the rocks – and getting one's hands dirty.

▲ A disused chalk quarry in Saffron Walden. Chalk was quarried in this area for making lime. This soft white limestone was laid down in a tropical sea at a time of great warmth when there was no ice at the poles.

fossils, in which case it might be dated between the layers above and below it, or by comparing it with an equivalent fossil-bearing bed elsewhere. The age of some types of rocks can be calculated by radiometric dating, which measures the proportions of radioactive elements present. This gives the absolute age of a rock. The dates of the geological periods are derived from many such measurements and comparisons.

Throughout its geological history, Essex has been a site of uplift and erosion with intervening periods of sagging and the deposition of new rock layers. Such changes result in an incomplete rock record, either because rock was not deposited, or because it was eroded later. Perhaps surprisingly, far more time is represented in the 'time gaps' than within the rock layers themselves; this is true for the rock record in most areas of the world. Nevertheless, even from this incomplete time record, we can now take a journey through the deep-time story of Essex rock. This book aims to reveal that story from the birth of the county 500 million years ago near the South Pole, through its gradual continental drift across the Earth's arid and tropical climate zones, until the past couple of million years or so of the current ice age. This ice age is a time of alternating temperate warmth and deep glacial cold, the time through which human beings have evolved. Some of them have occasionally led their lives across Doggerland – the area of the present North Sea – and up along the rivers of Essex.

The oldest rock to be seen at the surface in Essex is the pure white Chalk, made of the lime-mud of the seabed 80 million years ago. The 50-million-year-old London Clay is familiar to gardeners and engineers across south and mid-Essex, its dual character providing the hard and shrunken land of summer or the soft and swelling clay of winter. Across the Essex chalk and clay landscape, the most recent layers were spread by ice age rivers, notably when the Thames flowed as a wide torrent through mid- and north Essex. Less than half a million years ago a vast ice sheet spread across much of the county, as far south as Hornchurch and Billericay, leaving behind a thick layer of debris, the glacial till or 'chalky boulder clay', across much of Essex. The till plateau of Essex now provides

▲ The retreating cliffs at Walton-on-the-Naze tell a geological story of tropical seas, shell banks, rivers and storms. This is such a valuable section through the rocks of Essex that it is designated as a geological Site of Special Scientific Interest. Here the dark grey London Clay is topped by the Red Crag, succeeded above by sand, clay and gravel beds of ancient rivers – the Thames and Medway – and topped by wind-blown sand and life-giving soil.

fertile farmland. This Anglian ice sheet diverted the Thames towards its present position along the south Essex border. The current temperate interglacial stage started only 11,700 years ago. It is very different from the many interglacials that came before: this one has brought the impact of 4 million humans living across the area described in this book.

The importance of Essex geology

Geodiversity, alongside biodiversity, is an integral part of the natural environment. Geodiversity means the variety of rocks, fossils, minerals, landforms and soil, and all the natural processes that shape the landscape. The only record of the deep history of our planet lies in the rocks and sediments of land and sea. Here, and only here, can we trace the cycles of change that have shaped the Earth in the past, and will continue to do so in the future. This is particularly true in Essex, where the record of climate change during the ice age is preserved in quarries and coastal cliffs. The record is unique and much of it is surprisingly fragile, as well as being challenging to unravel in detail.

The rocks of Essex provide the obvious benefits of mineral resources such as sand, gravel, chalk and clay. Rocks are the foundation for the shaping of the landscape; they are also the basis of all farmland and food – and of natural soils which, in turn, regulate plant life, wildlife habitats and species. Geodiversity also has a cultural role to play, affecting the character of our built environment through the use of building materials and providing a sense of place across the varied areas of the countryside. An appreciation of geodiversity unveils these links between geology and landscape, nature and people, our perception of time – and the many changes to come.

▼ Scientists of all ages can investigate the rocks and fossils of Essex. Here a young collector is sieving the Red Crag sand at Walton-on-the-Naze for fossils.

Geoconservation involves the preservation of geologically important sites. Some sites can be given a measure of protection and the information and rock record used for education and research. Such sites are delineated as LoGS, Local Geological Sites, registered with local authorities throughout the county and protected – for instance from being covered over or destroyed – through the planning system. At a higher level of protection, geological Sites of Special Scientific Interest (SSSIs) are of ever-increasing importance as conserved 'witness sites'. In particular, SSSIs in the ancient Thames and other ice age deposits of Essex form an internationally important record, providing an opportunity for further geological research.

Over geological time the rocks and landscape of Essex are merely transient features, eroding rapidly into the sea. We as observers are here for an even more transient episode, during which we have the opportunity to enjoy the insight provided by observers, collectors, historians, archaeologists, geographers and geologists. They have built up a complex, ever-incomplete story of enormous changes across the county. As the impacts of current climate change heighten, we can now look back over the geological record of Essex, of its many climatic variations, and enquire into what is happening right now to this planet and its life… and into what may happen in the future.

Geological evidence revealed in Thorndon Country ▶ Park near Brentwood. Here you can see a cliff of gravel laid down by torrents of meltwater from an ice sheet that was situated just a short distance north of here.

2

Ice age gravels revealed by cliff regrading works at Holland-on-Sea in 2018. These gravels were laid down by a great river formed by the combined Thames and Medway.

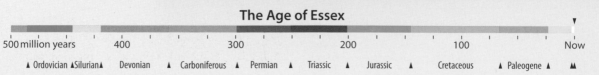

The Age of Essex

500 million years   400   300   200   100   Now

▲ Ordovician ▲Silurian▲   Devonian   ▲ Carboniferous ▲   Permian ▲   Triassic   ▲   Jurassic   ▲   Cretaceous   ▲ Paleogene ▲ ▲▲

The rocks of Essex

Seeing Essex rocks

Perhaps you have looked at the scenery while travelling around Essex and wondered why its 'lumps and bumps' and its flat areas are where they are; or maybe you notice a hill or slope as you cycle or walk. There is something to spot wherever you go, regardless of farmland, woodland or urban cover.

Sarsen stones by a farm track ►
near Gestingthorpe.

◄ Flint and quartzite in a cobble wall, Little Baddow church. Ancient walls were constructed of whatever rocks were available locally and can therefore tell us much about the local geology.

London Clay exposed ►
on the foreshore at
Walton-on-the-Naze.
It is soft – but it is
Essex rock.

When you are out walking, maybe in a country park or along a footpath, do you see the colour of the soil, a selection of pebbles, an area where the soil has worn away and different colours of sediment are revealed? Perhaps you notice 'odd rocks' on village greens, at crossroads or by trackways; or you spy an ancient church and wonder what the walls are made of. What are those layers of soft rock and fossils at the seaside? How old are they? Such observations help to reveal the deep history of this county and indeed anywhere we might find ourselves.

A footpath worn into the geology beneath: pale, ► sandy Claygate Beds at Thorndon Country Park.

The land beneath the soil: geological maps

The surface geology of much of Essex consists of a thin veneer of sands, gravels and clays, together called 'superficial' deposits. They have been left behind by rivers, ice and storms during the ice age. These layers, rarely much thicker than 40 metres in total, could be regarded as the icing on the geological cake – a good analogy, as they contain some remarkable fossils and exotic rocks yielding evidence of our most recent geological past. These superficial layers, shown in the map below, underlie wide areas of farmland and urban sprawl, yet river erosion is carrying them away to the sea, leaving a messy-looking picture with varied scenery. The map opposite reveals the layers that would be exposed if all the superficial deposits were removed. This shows the pattern of older rocks across Essex.

The geology of Essex has been mapped by the British Geological Survey (BGS) over many decades. Their maps show the rocks that occur at the surface. Maps, borehole records and much else can be accessed from the BGS website. Paper maps are also available for Essex at a scale of 1:50,000. With these resources it is possible to predict which types of rock, and therefore fossils, might be found at any place in Essex. They also help provide an insight into the rocks occurring at depth.

▼ Geological map of Essex – 'superficial'. These coloured areas show where the ice age superficial layers lie.

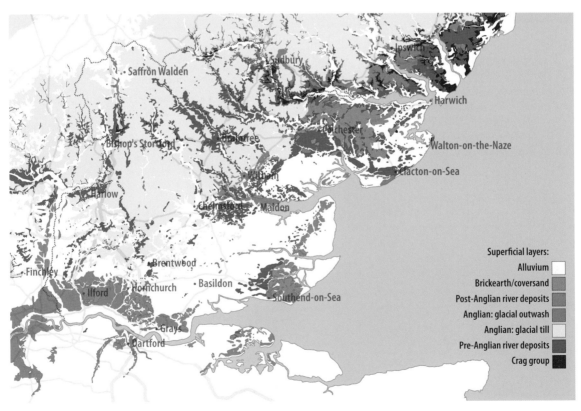

Superficial layers:
Alluvium
Brickearth/coversand
Post-Anglian river deposits
Anglian: glacial outwash
Anglian: glacial till
Pre-Anglian river deposits
Crag group

▲ Chalk bedrock seen in a quarry face
   at Chafford near Grays in south Essex.

▼ Geological map of Essex – 'bedrock'.
   These coloured areas show the underlying layers,
   all laid down before the ice age.

Bedrock layers:
Bagshot Sand
Claygate Beds
London Clay
Lower London Tertiaries
Chalk
Rocks older than Chalk

The shaping of the Essex landscape

The surface features of the landscape are largely controlled by the underlying geology. Although anyone might notice a change from the rolling countryside of north-west Essex to the flat coastal areas in the east, there are distinct types of landscape across the whole county: the Chalk areas, the glacial till plateau, the London Clay vale, the Claygate Beds and Bagshot Sand ridges and hills with high-level gravel hilltops, an enclosed bowl of fenland, the broad southern Thames terraces, and the coastal areas and river estuaries. Each landscape area has its own distinct character and the underlying rocks influence the vegetation, slopes, wildlife, land use, building and infrastructure, even the architecture.

The rolling chalk landscape of north-west Essex. ►
The hills of this part of Essex are an extension of the Chiltern Hills to the west. Here, bedrock is making the landscape.

▲ East Horndon church nestled within the Essex landscape.

Landscape types

The types of landscape across Essex have been assessed to help planners establish policies for conservation and development. A landscape map shows geographical areas each with a characteristic pattern of landscape. When you compare the geological map, which shows which rock layers are at the surface, with the Essex landscape character map, you can appreciate the strong influence of the geology upon the landscape.

Essex landscape character areas. ►
Based on Essex County Council information

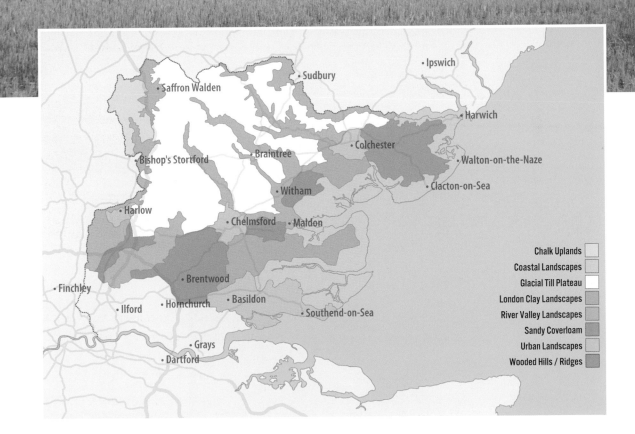

Chalk Uplands

Coastal Landscapes

Glacial Till Plateau

London Clay Landscapes

River Valley Landscapes

Sandy Coverloam

Urban Landscapes

Wooded Hills / Ridges

Ipswich

Sudbury

Saffron Walden

Harwich

Bishop's Stortford

Braintree

Colchester

Walton-on-the-Naze

Witham

Clacton-on-Sea

Harlow

Chelmsford · Maldon

Brentwood

Finchley

Hornchurch · Basildon

Ilford

Southend-on-Sea

Grays

Dartford

Digging into Essex

The soft sedimentary rocks of Essex are easily disintegrated by rain and frost and become covered with vegetation. Consequently, the only natural rock exposures to be found in Essex are on the coast, where the sea is continually eroding the rocks. Inland, it is only when the blanket of vegetation is removed, for example when a pit is dug, a major new road is constructed or a quarry is opened, that a window into the past is created. Any excavation that is sufficiently deep to penetrate the topsoil will expose the rocks beneath.

▲ Essex rocks beneath the landscape: the gravel bed of the former River Thames revealed at Birch Quarry near Colchester.

◄ Digging a pit to investigate ice age layers near the clifftop at Walton-on-the-Naze.

Dig down a third of a kilometre (1,000 ft)

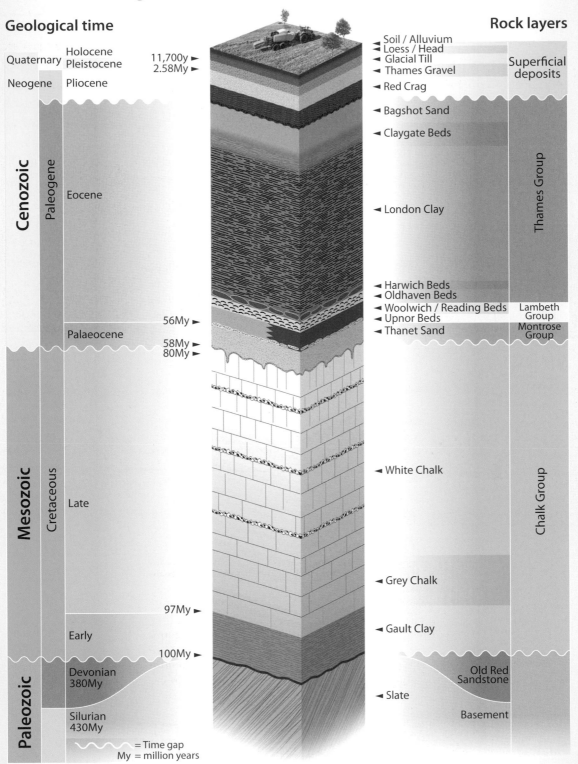

Geological time

Quaternary — Holocene
Pleistocene — 11,700y ►
2.58My ►

Neogene — Pliocene

Cenozoic

Paleogene — Eocene

Palaeocene — 56My ►
58My ►
80My ►

Mesozoic — Cretaceous — Late

97My ►
100My ►

Paleozoic — Devonian 380My

Silurian 430My

∿∿ = Time gap
My = million years

Rock layers

◄ Soil / Alluvium
◄ Loess / Head
◄ Glacial Till
◄ Thames Gravel
◄ Red Crag

Superficial deposits

◄ Bagshot Sand
◄ Claygate Beds

◄ London Clay

Thames Group

◄ Harwich Beds
◄ Oldhaven Beds
◄ Woolwich / Reading Beds — Lambeth Group
◄ Upnor Beds
◄ Thanet Sand — Montrose Group

◄ White Chalk

◄ Grey Chalk

Chalk Group

◄ Gault Clay

◄ Slate

Old Red Sandstone

Basement

Naming rock layers: decoding the jargon

The sediments laid down in the past may have hardened (e.g. into limestone) or remain relatively soft (e.g. gravel, clay), but these are all called 'rock' by geologists. Sedimentary rocks are usually deposited in layers or strata. Sometimes a layer is referred to informally as a 'bed'. To build up a more complete geological picture, the layers need to be correlated from place to place. The description, definition and naming of rock layers is termed lithostratigraphy (rock stratigraphy). The layers can also be described in other ways depending on the types of information available; for example, in biostratigraphy (life stratigraphy) fossils are used for correlation and in chronostratigraphy (time stratigraphy) the age of the rock is used.

Lithostratigraphy is fundamental to most geological studies. Sets of rock strata are assigned into units so that they are easier to talk about and to show on geological maps. For instance, the Woolwich Formation are a set of sandy, shelly and clay layers that are put together into one unit.

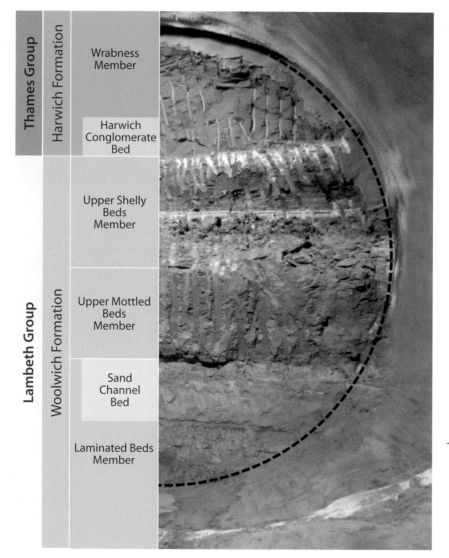

Thames Group	Harwich Formation	Wrabness Member
		Harwich Conglomerate Bed
Lambeth Group	Woolwich Formation	Upper Shelly Beds Member
		Upper Mottled Beds Member
		Sand Channel Bed
		Laminated Beds Member

◄ Rock layers in the Woolwich Formation revealed in the Thames Tideway Project, a super sewer constructed beneath London.
Tim Newman

The relationship of each unit in the whole rock record is set within a formal hierarchy: Supergroup, Group, Formation, Member and Bed. The layers of the Woolwich beds, for instance, are together called the Woolwich Formation. This, in turn, is part of the Lambeth Group. The units are usually named after a geographical locality, typically the place where exposures were first described. Another rock layer in Essex is referred to as the Thanet Sand Formation, even though Thanet is an area in Kent. A layer with a place-name keeps the same name wherever it occurs, under Essex or elsewhere.

The Formation is the basic rock unit for mapping purposes, as in the London Clay Formation. A Group is an assemblage of related and adjacent Formations – so the London Clay Formation is part of the Thames Group. A Member is a subdivision of a Formation, for instance the Claygate Member is part of the London Clay Formation. The terms 'Bed' and 'Band' are very often used in an informal way to aid description, such as with the Bullhead Bed and the Harwich Stone Band.

For some sedimentary rocks a useful correlation is one based on the fossils they contain – the biostratigraphy method. Because animals evolve over time, the species present may be different in each layer. A rock layer containing the same fossil species is of the same age wherever it is found, even if it has a different lithology (rock type) in different locations. Therefore, biostratigraphy is extremely valuable, sometimes enabling correlation of rocks across great distances. The formal units used are Eon, Era, Period, Epoch, Stage, Biozone and Bed. This approach is also particularly useful where the rock type is broadly similar over a long time period, such as during the Upper Cretaceous when chalk was the dominant rock type; for example, the *Micraster coranguinum* (a type of sea urchin) zone enables a particular layer of Chalk to be correlated in a thick and confusingly uniform pile of layers.

▲ The sea urchin *Micraster coranguinum* is a zone fossil in the Chalk.
Richard Hubbard

Age dating of rocks

These two methods of correlation, lithostratigraphy and biostratigraphy, provide only the *relative* ages for rocks and fossils; they merely help us put the rock layers into a time order, but without us knowing how old they actually are. The absolute dating of rocks, to give an age in millions of years, depends on the occurrence of radioactive isotopes that decay at a known rate over time. However, such isotopes are not always present in a particular rock and other methods have to be used in combination. Palaeomagnetism may be used with rocks that contain magnetic minerals – layers may be correlated using their 'locked-in' record of the direction of the Earth's magnetic field when the rock formed.

Tephrostratigraphy, the correlation of volcanic ash layers, has enabled the phases of rifting during the opening of the North Atlantic Ocean to be linked to ash layers in rocks across the north-west European continental shelf.

In recent years, major revisions of the naming and grouping of rock units have taken place. This is due to a more rigorous approach, with international collaboration and a much greater knowledge of the rocks themselves. Subsurface data from oil and gas exploration offshore and large-scale civil engineering projects have helped considerably. Names first introduced by Sir Joseph Prestwich in the nineteenth century have been more specifically assigned and new names have been introduced to accord with national and international standards. So, the names familiar to those of us who have been studying Essex geology for more than 30 years have been added to and changed. The study of geology is an ever-evolving scene. What a rock layer is called should not be a barrier to discovering how it was formed and the contribution it makes to the understanding of the story of our county through deep time.

▼ Volcanic ash in pale bands within the Harwich Formation at Wrabness.

▲ Coastal erosion reveals an ice age deposit at Wrabness. Correlation is based on the fossils it contains.

Bedrock:

London Clay
Lower London Tertiaries
Chalk

Superficial layers:

Alluvium – clay, silt, sand & gravel
Solifluction head – silt, sand & gravel
River terrace – sand & gravel
Anglian glacial outwash – clay, silt, sand & gravel
Anglian glacial till
Kesgrave Thames sand & gravel

• Gestingthorpe

• Bulmer Brick & Tile Co.

◄ A geological map of a small area of Essex near Sudbury, showing the bedrock and the superficial rock layers that appear at the surface. All types of stratigraphy are used in making and revising such maps.

Time gaps

Sediments are not deposited continuously over geological time. The rocks we see were formed in various geological settings. The potential for sediments to be preserved and then turned into rock depends on the type of environment, such as whether the area was sinking and a sediment layer was quickly buried by the next influx of material, rather than being washed away. For large expanses of geological time there may be no record in the sediments. Instead there will be time gaps. The rock layers beneath Essex illustrate this remarkably well.

The larger time gaps usually represent periods when an area was above sea level and the land was worn away by erosion. Other time gaps were caused when sediment was not laid down or preserved. The next layer of sediment then covered an area where the rock record was missing, leaving a time gap.

Essex lies within an ever-changing edgeland of the European continent. It was 'separated' from mainland Europe whenever a shallow sea occupied the area of the subsiding North Sea Basin. As a consequence, over millions of years, the surface of Essex has fluctuated between low land and shallow sea. Thus, sedimentary layers have only occasionally accumulated and some of these layers have been eroded away subsequently, resulting in a discontinuous rock record with many time gaps.

The layers of rock beneath Essex are shown ▶ on this timescale to reveal the time gaps.

▼ Geological research has revealed that there is a 50-million-year time gap between this dark grey London Clay and the overlying shelly Red Crag layer along the cliffs at Walton-on-the-Naze.

Time Gap

More gap than record: Essex rocks in geological time

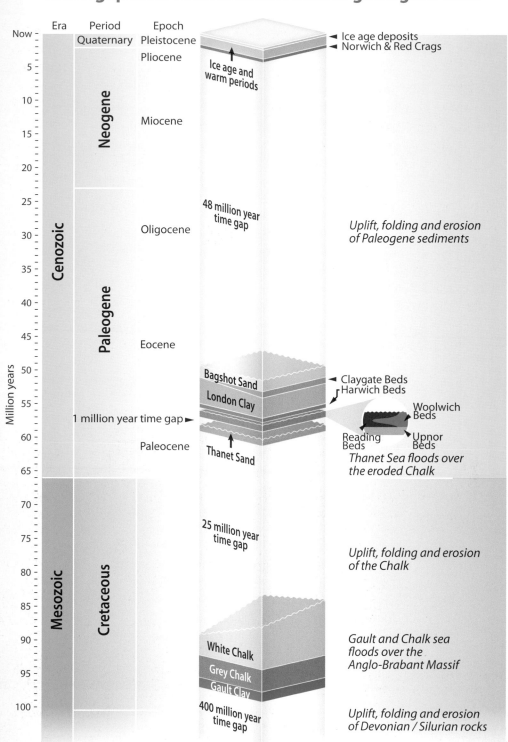

3

The birth of Essex close to the South Pole 500 million years ago – on the seabed around a chain of volcanic islands.
Wikimedia Commons /R.G. Spear

Deep History

500 million years	400	300	200	100	Now
Origins of Essex near the South Pole	Foundations of Essex	Essex crosses the Equator	Dinosaurs in Essex	Chalk Sea	Global cooling

The deep history of Essex

Hidden history: the oldest rocks beneath Essex

Compared with many other parts of Britain, even the bedrock geology of Essex is young in geological terms. At the surface, the oldest rock exposed in Essex is the Chalk, which was formed about 80 million years ago. We can see chalk hills and quarries in north-west Essex and also in the south near Purfleet; but what was Essex like before the Chalk was formed? The answer is found in the rocks deep beneath the county which are revealed in samples from boreholes sunk at various times in search of water and coal. Where there are gaps in the Essex rock record, the study of rocks exposed in other parts of Britain help to complete the story.

If you could dig down about 300 m (1,000 ft) beneath the surface, you would come to the oldest rocks in our story of Essex. These are between 440 and 360 million years old. They were deposited in the Silurian and Devonian Periods when the map of the continents and the climate in Essex were very different from today. The deep history of Essex over the past

500 million years is a fascinating tale of volcanoes, huge mountain ranges, deserts and tropical forests, culminating in dinosaurs tramping over a large island before the land was finally drowned beneath the sea.

Around 500 million years ago, before the end of the Cambrian Period, the area that was to become Essex was covered by a shallow sea, with sediments spreading from a nearby volcanic island chain. The area of England and Wales was near to the south pole. It lay at the margin of a very large continent that geologists have called Gondwanaland.

The volcanic activity continued as part of this continent split away, drifting northward as a small continental fragment called Avalonia – England and Wales were at its eastern end. Evidence for these movements comes largely from magnetic measurements in rocks that show where the poles were when that particular rock was formed, as well as from the climate zones shown by fossils and the nature of the

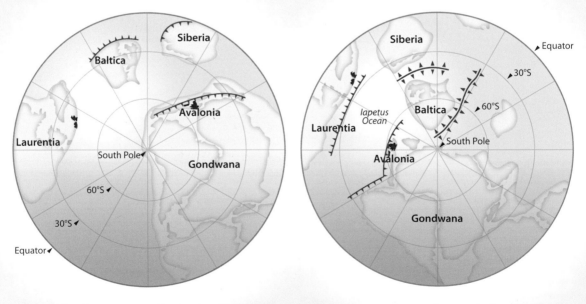

▲ 500–550 million years ago: England and Wales, with Essex, are not far from the South Pole and are part of a shallow seaway near a string of volcanoes along the edge of the Gondwana continent.

▲ 490 million years ago: a fragment of continent – 'Avalonia' – splits from Gondwana and drifts away, carrying England and Wales with it.

sediments themselves. The Earth's tectonic activity continually splits, moves and merges the continents.

The map of continents 440 million years ago shows that Essex is still part of Eastern Avalonia, by then a volcanic island chain in the middle of the Iapetus Ocean. Huge rivers deposited sands, silt and plenty of mud into the surrounding sea. We have fossil evidence of brachiopods and trilobites that lived on the seabed and colonial graptolites that floated above. Avalonia drifted slowly north, nearing the tropics and heading towards the large continents of Baltica and Laurentia.

By 420 million years ago Avalonia, including England and Wales, collided first with Baltica and then with Laurentia which included the area of Scotland. A huge mountain chain was created along the collision zone, much like today's Himalayas. The thick layers of Silurian mudstones were squeezed and converted into the hard rocks discovered in the deep boreholes at Weeley and Harwich beneath Essex and also at Stutton just over the border in Suffolk.

◄ 440 million years ago: Iapetus Ocean is closing and Avalonia heads for collision.

▼ Cross-section of the Earth showing the plate tectonic drift of Essex on the island continent of Avalonia 440 million years ago.

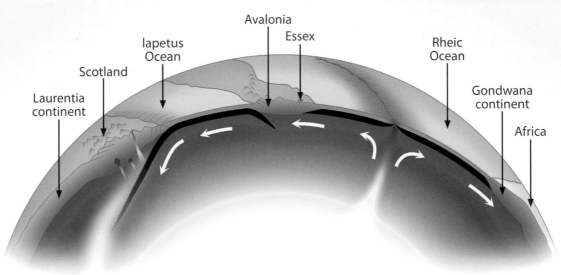

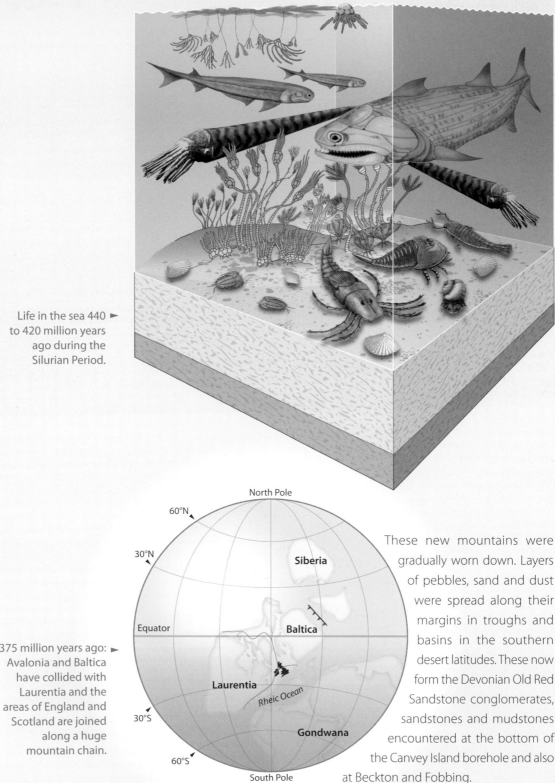

Life in the sea 440 ► to 420 million years ago during the Silurian Period.

375 million years ago: ► Avalonia and Baltica have collided with Laurentia and the areas of England and Scotland are joined along a huge mountain chain.

North Pole

60°N

30°N

Siberia

Equator

Baltica

Laurentia

Rheic Ocean

30°S

Gondwana

60°S

South Pole

These new mountains were gradually worn down. Layers of pebbles, sand and dust were spread along their margins in troughs and basins in the southern desert latitudes. These now form the Devonian Old Red Sandstone conglomerates, sandstones and mudstones encountered at the bottom of the Canvey Island borehole and also at Beckton and Fobbing.

▲ A vision of Essex during the Devonian Period 375 million years ago when the Old Red Sandstone was deposited.
Wikimedia Commons/Gyik Toma

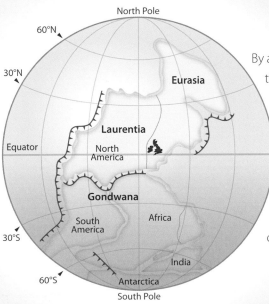

300 million years ago: ► the Rheic Ocean has closed and Gondwana has collided with the northern continents – tropical Britain is within the new Pangea supercontinent and Essex has crossed the equator.

By around 320 million years ago the Rheic Ocean finally closed as the northwards-drifting Gondwanaland continent collided with Laurentia and Eurasia, forming yet more mountain chains. Essex was set in place within one vast new super-continent known as Pangea.

▲ Tropical coal forest may have covered Essex during the Carboniferous Period 300 million years ago, but no coal has been detected beneath Essex.
Wikimedia Commons/Lubasi

The new Pangea continent continued to drift slowly northwards. By 310 million years ago Britain straddled the equator and the climate was humid, much like the Amazon and the Congo today. The uplands of Essex were flanked by forests and swamps. The coal in the British coalfields was formed at this time, including the Kent coalfield – for which there is no evidence at the surface. In an attempt to find more coal, deep boreholes were drilled across East Anglia in the 1890s and again in 1953 on Canvey Island, but this drilling revealed no coal. It is an interesting thought that, had the borehole results been different, Canvey Island might have become a coal-mining town; this is an example of how geology affects the character of the landscape.

The geological structures – folds and faults – formed during all these mountain-building events are gradually becoming better understood. Evidence comes from deep boreholes, geophysics and correlation of rocks across Britain and beneath Essex. Repeated movements along the deep-seated faults in the Earth's crust still result in earthquakes under Essex and surrounding areas. Through many years

▲ The arid uplands of Essex during the Triassic Period 240 million years ago were eroded over millions of years and red dune sands were spread across the area as the hills were levelled – they would have looked similar to these in the Namib desert. *Wikimedia Commons/Robur.q*

of detailed research, a timetable of events has been constructed. It is being fine-tuned continually, but the deep-time story of Essex is becoming ever clearer.

As the landmass of Pangea – including the area of Essex – continued to drift further northwards, it entered the northern tropical desert zone at the same sort of latitude as the Sahara Desert today. Through Permian and Triassic times, 300 to 200 million years ago, the climate was once again arid. Although sand dunes and salt-lake deposits were laid down in other parts of the British Isles at this time, no trace of these is found in the layers of rock beneath Essex. Here, the hills were continually eroded and the dune sands and the wadi gravels were eventually transported away from the area.

The upland area persisted for hundreds of millions of years as the collisions of the continents created an extra-thick 'platform' of the Earth's crust under Essex. Stretching from the English Midlands across to Belgium, this thicker piece of crust is called the 'Anglo-Brabant Massif'.

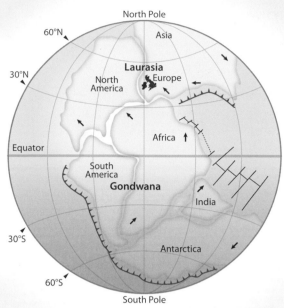

▲ 240 million years ago: Essex is in a vast continental desert while the Tethys Ocean opens out to the east.

▲ 195 million years ago: shallow seas start to flood the continent across the area of Britain. Essex, on a 'platform' of thicker crust, remains an island in the new seas.

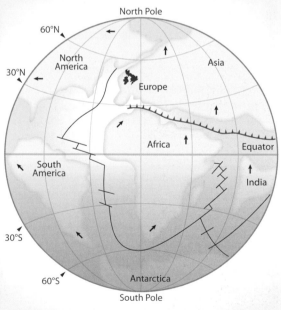

▲ A 140-million-year-old fossil tree from Portland, Dorset, now at Thorndon Country Park near Brentwood. It was similar to the pine trees that now grow close by – but the dinosaurs no longer lurk here. The Jurassic island on which these trees grew is now buried far beneath Essex.

▲ At the end of the Cretaceous period, 66 million years ago, Britain was caught between the opening Atlantic Ocean and the closing Tethys Ocean where the collision of Africa with Europe was beginning to form the Alpine mountain chain. The progress of Essex geology was – and still is – influenced by both of these prolonged plate tectonic episodes.

Dinosaur island

Eventually, most of the desert area became inundated by a shallow sea linked to the Tethys Ocean which had opened up to the east. In this sea swam the plesiosaurs and ichthyosaurs of the Jurassic Period, along with ammonites and other familiar Jurassic animals. Fossils of these are found in the marine sediments of the Dorset coast, but Essex continued to be above the waves. Dinosaurs would have roamed on this 'Essex Dinosaur Island', which lay along the ancient platform of the Anglo-Brabant Massif stretching across to the area that is now Germany. We can imagine megalosaurs, allosaurs, stegosaurs and diplodocus walking the pine-forested land, with archaeopteryx gliding overhead.

By the time of the great T. rex, the Cretaceous sea was lapping onto Essex Island. The mid-ocean ridge between Europe and America started to rise, causing the ocean waters to spread across the lands of the continental margins, forming very extensive seas. By 100 million years ago a cloak of sediment finally spread across the old Jurassic island, concealing its rocks. The Anglo-Brabant Massif, that thick platform of crust with its surface of ancient uplands, was at last buried under new rock layers. Yet the influence of this astonishingly persistent platform is still in evidence even in the later, overlying rock layers of Essex.

▼ Essex Dinosaur Island lapped by the Jurassic sea 165 million years ago.

4

Deep beneath Essex, the Earth's crust forms a tough and resistant 'platform' about 38 km (24 miles) thick. Stretching from Belgium to the English Midlands, it is known as the Anglo-Brabant Massif. The earliest samples of 420-million-year-old rock from the top of the Massif were found beneath Essex at Harwich in 1857. Later, a core of slate was recovered in 1896 from the deep borehole at Weeley.
Gerald Lucy / Colchester Natural History Museum

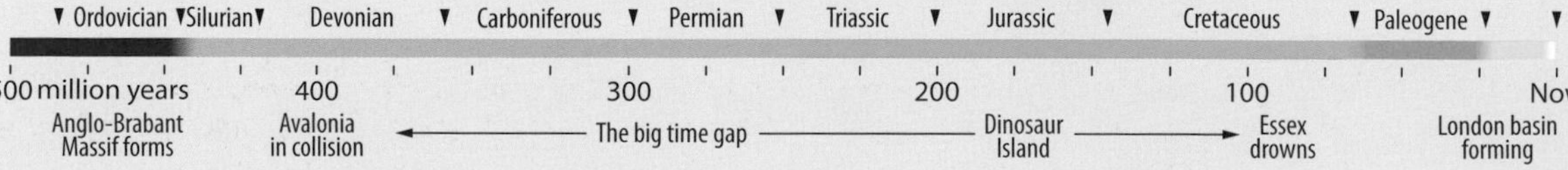

▼ Ordovician ▼Silurian▼ Devonian ▼ Carboniferous ▼ Permian ▼ Triassic ▼ Jurassic ▼ Cretaceous ▼ Paleogene ▼ ▼

| 500 million years | 400 | 300 | 200 | 100 | Now |

Anglo-Brabant Massif forms   Avalonia in collision   ←——— The big time gap ———→   Dinosaur Island ———→   Essex drowns   London basin forming

WEELEY
1150 FEET

Anglo-Brabant Massif

Weeley •

The geological structure of Essex

Down to the basement

We may not think too much about what is beneath our feet, but the accumulation and shifting of rocks deep down has had a great effect upon the land surface and consequently upon our lives today. In Chapter 2 we showed the rock layers that you would discover if you were able to dig down into Essex for about one-third of a kilometre (1,000 ft). Down to that depth the rocks are in fairly even layers of relatively soft rocks such as clay, sandstone and chalk. Then suddenly the digging becomes much more difficult – the rocks beneath are far older and tougher. At this depth is a segment of thicker, stronger crust, a platform of resistant rock. This platform is known as the 'Anglo-Brabant Massif'. This massif makes up most of the thickness of the crust under Essex; it is thicker here than in surrounding areas, about 38 km down to the boundary with the mantle – the distance from Chelmsford to Colchester.

The origins of the Anglo-Brabant Massif go back more than 500 million years, to when this part of the crust was a small fragment of an ancient super-continent, Gondwanaland (discussed in Chapter 3). Eventually it became involved in a series of continental collisions: the continental fragment of Avalonia collided with Baltica around 430 million years ago, and then with Laurentia. As the Rheic Ocean expanded from the south, most of the rocks that now make up the Massif were compressed, becoming hardened and transformed by heat and pressure (metamorphosed), and folded into mountains. This resulted in a thicker and more rigid crust than in surrounding areas. Ever since these continental collisions, the Anglo-Brabant Massif has remained a significant feature, influencing the deposition of later rock layers.

Detailed studies of the geology of the Massif are hampered by the lack of penetrating boreholes, plus the fact that those boreholes that do exist were largely drilled in the 1890s before the advent of modern analytical methods. The sections of core retained by various museums and other organisations are the sole record, making their secure curation particularly important.

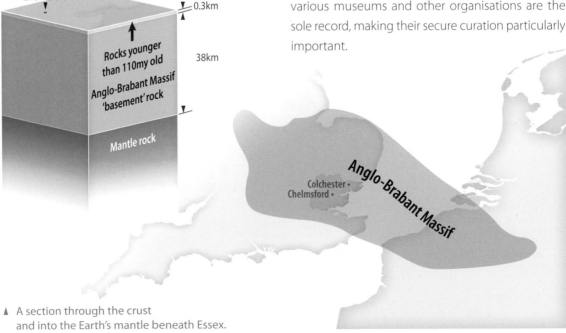

▲ A section through the crust
and into the Earth's mantle beneath Essex.

Discovering the deep rocks beneath Essex

In the nineteenth century, wells and boreholes penetrating beneath the Chalk were largely drilled as the result of speculation. Economic resources such as water, coal and iron ore were in great demand during the Industrial Revolution as the population increased. Geologists and speculators made comparisons with geological sequences described in other parts of Britain in the hope of predicting the occurrence of further supplies of economic minerals.

The first well to penetrate strata below the Chalk in Essex was dug at Harwich by the forerunner of the Tendring Hundred Water Company. The lack of good drinking water had long been a complaint among residents, so despite the failure to find water with two previous wells, the sinking of an ambitious new borehole was commenced in 1854. Two years later the Harwich borehole had penetrated deep into the Chalk but, surprisingly, had still not encountered a satisfactory supply of water and the decision was made to continue through the Chalk to the strata below.

In November 1857, three years after the work had begun, the borehole had been drilled right through the Chalk and the Upper Greensand and the Gault, into hard, slate-like rock. Although no satisfactory water supply was obtained from this borehole, and it was to be several more years before a good supply was found for the town, the borehole had reached a depth of over 300 m (1,000 ft) and proved to be of great value to science. The hard basement rocks of Essex – the Anglo-Brabant Massif – had been revealed for the first time. The discovery site was by the harbour, near the pier, just west of the Great Eastern Hotel. Very few further deep boreholes have been drilled in the region, most of these in the search for coal.

The first borehole ► to penetrate the deep basement of Essex was drilled just to the right of the later Great Eastern Hotel at Harwich. The log reveals the discovery of the rock of the Anglo-Brabant Massif at the bottom of the well.

184   ESSEX WATER SUPPLY.

Harwich, cont.

5. By the Harbour, just west of the Great Eastern Hotel. 1854–7.

PRESTWICH, *Quart. Journ. Geol. Soc.*, vol. xiv, p. 249; and BRUFF, *Proc. Inst. Civ. Eng.*, vol. xix, p. 21.

6 ft. above high-water mark.

Shallow shaft, the rest bored.

	Thickness. Ft.	Depth. Ft.
Earth [made ground, mud ?]	10	10
Red gravel	15	25
London Clay. Platimore, mixed with chalk and white sand [? Blackheath Beds in part]	23	48
[Blackheath Beds.] Coarse dark gravel	10	58
[Reading Beds, 20 ft.] Plastic clay	7	65
Bluish plastic clay, with green sand (1)	1½	66½
Red plastic clay, with green sand (1)	3½	70
Greenish sand (1)	2	72
Greenish and red sand (1)	3	75
Dark red (or blue) clay	3	78
[Upper, Middle and Lower Chalk, 890 ft.] Chalk, with flints in layers 5 or 6 ft. apart, and with shells	690	768
Chalk without flints (2)	162	930
Chalk Marl with thin layers of rocky chalk	38	968
Gault mixed with green sand	22	990
Gault	39	1029
Hard dark bluish-grey slaty rock (3) [some of the specimens of this seem to show planes of bedding, cleavage, and jointing]	69	1098

The quest for coal, the Channel Tunnel and the deep rocks of Essex

Sir Joseph Prestwich, who had written much about the geology of southern Britain, including work in Essex, had a seat on the Royal Commission on Water Supply. He was also appointed to the Royal Coal Commission from 1866 to 1871, which concluded with the Report on the Probabilities of finding Coal in the South of England (1871). Coincidentally, in 1880, work began at Dover's Shakespeare Cliff on a Channel Tunnel from Dover to Calais in which Prestwich was also involved, but in 1882 the government halted the work while it considered the military implications. With its workers lying idle, the Channel Tunnel Company decided to drill boreholes to investigate the geology of Kent. The result was the discovery of both iron ore and coal in 1890. What was found in Kent could possibly occur in Essex, so the Eastern Counties Coal Boring and Development Syndicate was formed to investigate further. In 1894–95 they sunk a borehole at Stutton on the Suffolk side of the Stour estuary, followed by a borehole at Weeley in Essex in 1896.

The Stutton borehole encountered 'broken and jointed grey shaly sandstone at 297 m (994 ft) below the surface'; 4.9 m (16 ft) below this were 'hard, bedded and cleaved rocks with a high dip'. Drilling continued to a depth of 464.9 m (1,525 ft). The Weeley borehole, drilled down-dip from Stutton, penetrated the same deep basement rocks beneath the Chalk and Lower Cretaceous rocks at a depth of over 330 m (1,100 ft) and continued to 372 m (1,221 ft) before the project was finally abandoned. As at Harwich and Stutton, the basement was found to be composed of Silurian rocks (approximately 420 million years old) and not the younger Carboniferous Coal Measures that had been hoped for.

A further borehole at Great Wakering near Southend was suggested but finance was not forthcoming. Directors and shareholders had already invested a considerable amount of money in the venture and the syndicate was finally wound up a year or so later. A section of the core from the bottom of the Weeley borehole is on display in Colchester's Natural History Museum and another has been preserved in the collection of the Essex Field Club. The site of the borehole is on the northern side of the stream about 100 metres north-east of Weeley railway station.

A core of 420 million ► year old slate from the Silurian period sampled from the 1894 deep borehole at Stutton.

The basement rock

The 'old rocks' sampled from boreholes into the Anglo-Brabant Massif are all described as 'hard, steeply dipping, grey sandy shale, cleaved and contorted'. They are thin, alternating layers of fine sand and mud, hardened and compressed over time. Graded bedding in these layers indicates that they were deposited in deep water from fast-moving seabed avalanches of sediment travelling from the continental edge.

The rocks contain telltale flakes of the grey-green mineral chlorite that had grown in the solid rock, parallel to the cleavage. They indicate the same low-grade metamorphism, pressure crystallisation, that affects many of the Silurian rocks seen at the surface in Wales – these were formed at about the same time.

▲ Silurian slate rock on the coast of Wales, similar to the Silurian basement rock found 300 m (1,000 ft) beneath Essex. *Wikimedia Commons/AliV*

Graded bedding, each layer deposited by a seabed ► avalanche, coarse sediment was followed by finer particles as the current waned.

The age of these rocks caused much debate at the time of drilling, especially when the fossil thought to be of *Posidonia*, a bivalve indicative of Carboniferous rocks, found in the core from Harwich, was later identified as merely a shell-like 'conchoidal' fracture. Whitaker of the Geological Survey, whose learned opinion was sought, recorded a chance break in a core sample that yielded a specimen of *Orthoceras*, a type of straight nautilus of Silurian age. After more than 150 metres (500 feet) of these hard, cleaved shales had been drilled in the Stutton borehole, he was asked whether Carboniferous rock might still be found. He advised that the younger Carboniferous rocks were only likely to be found if there was a thrust fault and, even if they were, their great depth would preclude economic exploitation. Other experts were consulted and the conclusion was that these rocks were of Silurian age (between 440 and 420 million years old). This has been confirmed by modern techniques.

Canvey Island borehole

Devonian rocks, slightly younger than the Silurian rocks found further north, were proved in the southern flank of the Anglo-Brabant Massif under Canvey Island. The Canvey Island deep borehole was drilled in 1953, and reported on by the BGS in 1964, as part of a renewed coal exploration programme. A gravity survey had previously shown an area of less dense rock which, it was thought, could be Coal Measures. Instead, over 130 m (430 ft) of older rocks were drilled and cored in the borehole. These rocks, of Old Red Sandstone (ORS) type, were encountered at 401.6 m (1,317.5 ft) at a fissured surface with phosphate nodules beneath the much younger Gault clay. The age of the ORS sediments was proved to be Lower Devonian (420 to 490 million years old) by the presence of fossil ostracods, small crustaceans with a bivalved shell that show rapid evolution. Fossils of early plants were also found. The borehole was the first in Britain in which the direction and dip of subsurface rocks were measured by the use of sophisticated instruments. The ORS strata were found to be dipping eastwards, towards the centre of the gravity anomaly that the borehole was drilled to investigate.

◄ Old Red Sandstone – 370 million year old conglomerate laid down in the Devonian Period in the Isle of Arran, Scotland. It is possible that similar rocks exist beneath Essex.

Geophysical mapping

Apart from borehole data, geophysical surveys are the only means of establishing the geology and structure of the rocks deep below the surface. Geophysical surveys measuring both gravity and magnetic fields have been carried out and digitally enhanced for the whole country but, due to that lack of economic targets deep under Essex, there has been no acquisition of costly seismic data here. Thus, there is no clear picture of the structures of rocks in, and just above, the Anglo-Brabant Massif beneath Essex. Determining the configuration and structure of the Massif, and hence its effect on the overlying rocks, is largely based on the less-definitive gravity and magnetic surveys.

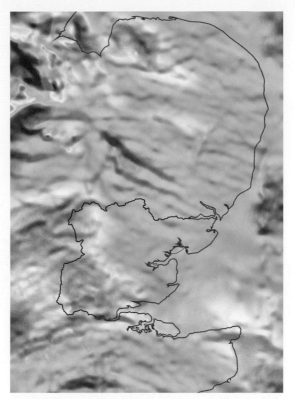

Magnetic anomaly map. ►
British Geological Survey © UKRI 2021

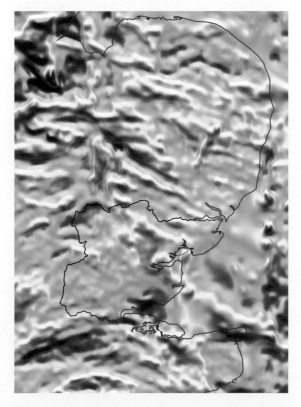

A gravity map is made by detecting small differences in the density of rocks and the thicknesses of each layer. These variations produce small changes in their gravitational effect. By measuring these effects with very sensitive instruments and displaying the data on maps, it becomes possible to spot subsurface variations. Similarly, by measuring very small changes in the magnetic field, the effects of different rock types can be detected. The magnetic field variation across Essex shows no particular features and calculations give the depth to 'magnetic basement' as greater than 5 km, a long way below the top of the Anglo-Brabant Massif. This indicates that there are no iron-bearing rocks that would have a magnetic signature at a shallow depth.

◄ Gravity anomaly map.
British Geological Survey © UKRI 2021

Underlying geology and structure

Gravity data, backed up by the few deep boreholes that have been drilled, give some idea of the geology within the Anglo-Brabant Massif itself. Much of the story is gathered from studies of the tectonic movements that affected rocks of similar age elsewhere, in Wales, the Lake District and Belgium, and also across the Atlantic.

The large north-west to south-east trending gravity features further north in East Anglia are thought to trace the subduction zones formed when Eastern Avalonia collided with Baltica 450 million years ago. This was the first of a complex series of events. These were followed by the collision with Laurentia when England and Wales joined with Scotland and then by the collision of further continental fragments across the Rheic Ocean from the south. The south-western part of the Anglo-Brabant Massif is composed of the remains of an even older resistant platform, the Midlands Microcraton, that influenced deposition in Silurian times 440–420 million years ago. The borehole at Ware in Hertfordshire found shelly Silurian beds formed in a shallow sea environment across this block of crust, whereas the Silurian beds found in the boreholes to the north-east of this block, in north Essex and Suffolk, are of deep-water deposits. Mapping of the cleavage directions in the rocks shows how the deformed mountain belts of Silurian rocks of the Welsh Basin and the East Anglian Basin were wrapped around this segment of stable crust.

All of the deep rock compression during mountain-building recrystallised the minerals in the rocks. The resulting changes to the rocks – metamorphism – depend on the degree of heating and squeezing. Studies from borehole cores in East Anglia have shown that the metamorphic grade within the folded Silurian rocks increases northwards, but is still of relatively low grade compared with, for instance, metamorphism seen in Scotland from earlier times. The rocks seen in the deep borehole cores are of slate grade, which is comparable with the slates from Wales, much used for roofing.

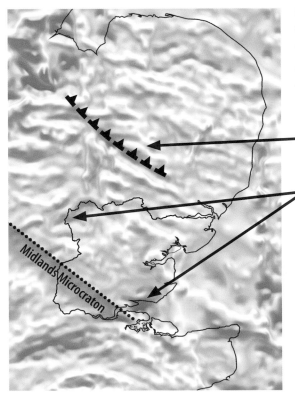

Glinton Thrust – a Caledonian subduction zone reactivated by later Alpine movements.

Gravity lows over areas now proven by boreholes to contain thick sequences of Devonian rocks such as at Canvey Island.

◄ The gravity anomaly map shows strong crustal structure with north-west to south-east trending lineations reflecting the geometry of the continental collisions during the formation of the Anglo Brabant Massif. The gravity lows, shown in blue, were once thought to be due to granite intrusions, but the drilling of boreholes has proved thick sediments of Devonian age.
British Geological Survey © UKRI 2021

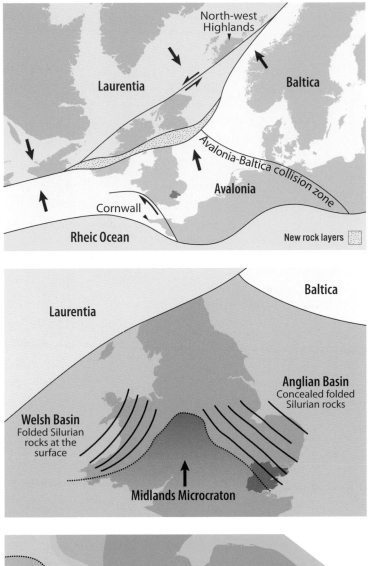

◄ Map of the continental collision zone between Avalonia, Baltica and Laurentia.
After Woodcock and Strachan

◄ The Silurian rocks above the Midlands Microcraton have shallow-water fossils; rocks of the Welsh and Anglian Basins are of deep-water sediments. Cleavage directions in Silurian slate reveal how the ancient crust of the Microcraton was pushed into the Welsh and Anglian rocks.

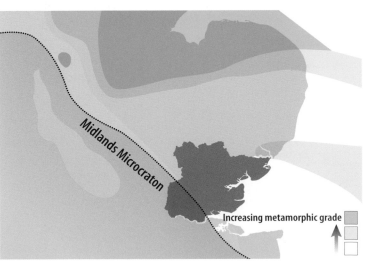

◄ Metamorphic grades in Silurian rocks of East Anglia. The colours reveal the degree of squashing of the rocks.

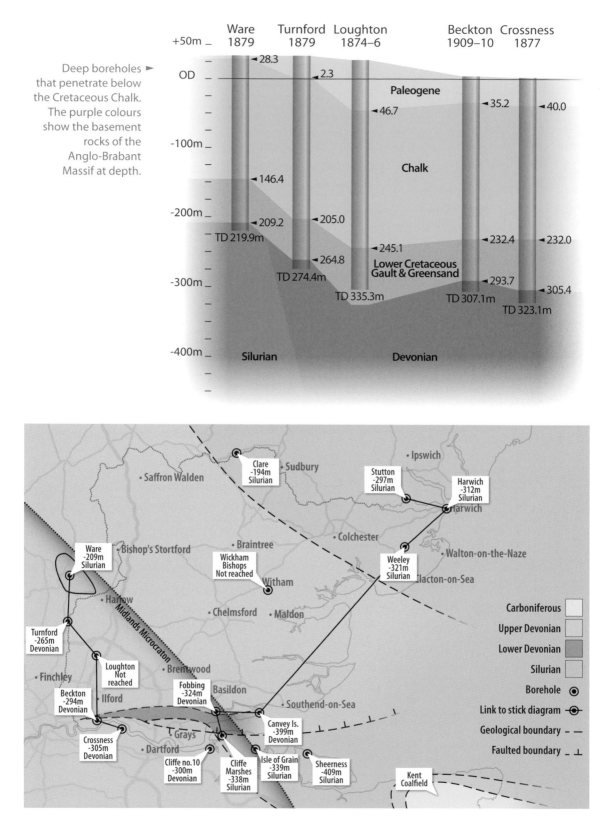

Deep boreholes ► that penetrate below the Cretaceous Chalk. The purple colours show the basement rocks of the Anglo-Brabant Massif at depth.

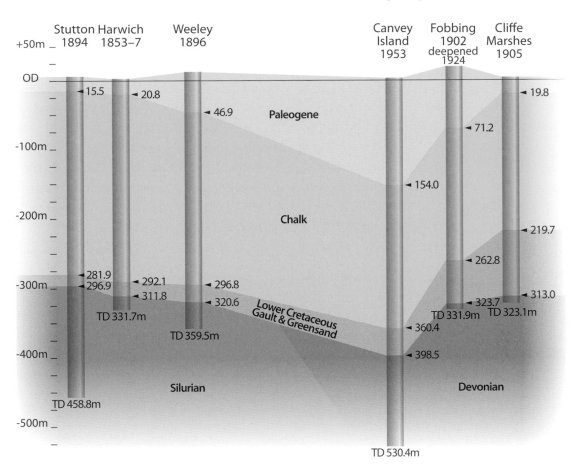

The Silurian mountains made of slate were rapidly eroded during Devonian times to produce thick layers of Old Red Sandstone, like that sampled in the Canvey Island borehole. By 360 million years ago, large amounts of Old Red Sandstone had accumulated. By comparing the area under Essex with better-exposed and explored areas to the west, it seems that large thicknesses of Devonian sediments must have been eroded. However, in mid-Essex some of these deposits may still be preserved, giving rise to a cluster of low gravity features. Only drilling would be able to prove this possibility.

From the end of Devonian times 360 million years ago, to late Cretaceous times 110 million years ago, there is no evidence in Essex for any sediments being deposited across the Anglo-Brabant Massif. The area continued to be uplifted as the Rheic Ocean closed and continents converged from the south. The Massif was a large island from then on. Its rocks were worn down, resulting in the 400-million-year time gap in the boreholes of north Essex. The Anglo-Brabant Massif was eventually flooded by a worldwide rise in sea level 110 million years ago. Plate movements related to the opening of the Atlantic to the west then became dominant.

◄ The geology at the buried surface of the Anglo-Brabant Massif basement of Essex, showing the few boreholes that have reached that far down.
After N.J.P. Smith, BGS Pre-Permian Geology map

A story of tilting and squeezing: the London Basin

The 'bedrock' seen on the geological maps of Essex is made of layers that were laid down in the last 110 million years. These lie across the top of the basement, the Anglo-Brabant Massif – the underlying platform of much older rocks. However, these overlying bedrock layers are no longer completely flat-lying, as they were when each layer was first deposited. Instead, the rocks underlying the whole of Essex are in the form of a shallow trough known as the 'London Basin'. External tectonic forces are responsible for this distortion in the pattern of Essex rock and these forces have been in action, in stages, through more than 60 million years.

The London Basin is edged by the scarp of the Chiltern Hills to the north-west and its extension along the chalk hills of Cambridgeshire and north Essex and, to the south, by the North Downs of Kent. The slopes of the rock layers dip inwards from these two edges to form a V-shaped basin, although there are minor irregularities caused by faulting and folding.

▲ Cambridgeshire, viewed here from the north-west Essex borderlands.

The tectonic forces

Two separate tectonic forces have resulted in the Basin's V-shape. The first was caused by tension in the crust to the west of Britain, when a rift valley system started to open out from around 60 million years ago. This split eventually widened to form the North Atlantic Ocean and a new, western edge of the European continent, of which Britain is a part. As the edges of the rift lifted, the western part of Britain was raised up and tilted to the east and south-east. Along with this gradual movement, the buried platform of the Anglo-Brabant Massif was tilted, although it was not itself distorted into a basin shape. The overlying bedrock layers of Essex were tilted along with the platform. Throughout those millions of years of tilting and erosion, the surface rock layers across England were worn down into a series of scarp-edged hills and vales. This pattern of erosion is remarkably well displayed as a set of prominent stripes seen across the geological map of the area to the west and north-west of Essex. The resulting landscapes form notable and often picturesque features of the English countryside such as the Chiltern hills and the Cotswolds.

The second tectonic force, around 40 million years ago, came with an episode of mountain-building as the African continent crushed into parts of the European continent. The south-eastern part of England was compressed while the Pyrenees mountain chain formed further south. However, the tough crust of the Anglo-Brabant Massif acted as a resistant block. To the south of Essex, the thick rock layers of the Weald of Kent and Sussex were squeezed against the southern edge of the Anglo-Brabant Massif and folded into a dome shape, hence the oval 'blister' form of the geography of the Weald. The rock layers of north Kent and south Essex are consequently tilted downwards to the north – they have a northerly 'dip'. The combined effects of the tilting down towards the south-east and the compressive tilting towards the north have, together, left a V-shaped 'trough' in the structure of the bedrock layers: the London Basin. The youngest bedrock layers are still preserved in the central part of the basin, while erosion has now exposed the Chalk at the northern and southern edges of Essex.

▲ The development of the London Basin from the time of the Chalk Sea (top), the later tilting of the land down towards the south-east (middle) and subsequent squeeze of the crust from the south, resulting in a V-shaped 'basin' across Essex.

▼ The geological bedrock map shows the structure of the London Basin. The oldest rocks are in the north and the south, with the youngest along the axis.

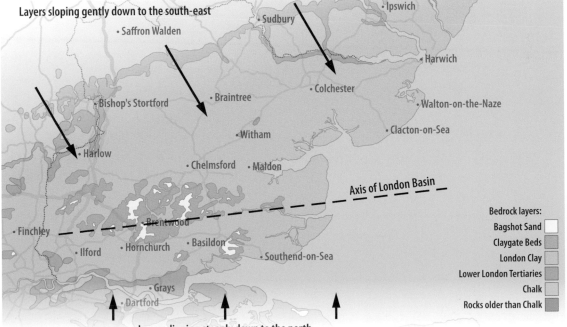

Layers sloping gently down to the south-east

- Ipswich
- Sudbury
- Saffron Walden
- Harwich
- Colchester
- Bishop's Stortford
- Braintree
- Walton-on-the-Naze
- Witham
- Clacton-on-Sea
- Harlow
- Chelmsford
- Maldon

Axis of London Basin

- Finchley
- Brentwood
- Hornchurch
- Basildon
- Ilford
- Southend-on-Sea
- Grays
- Dartford

Layers dipping steeply down to the north

Bedrock layers:
Bagshot Sand
Claygate Beds
London Clay
Lower London Tertiaries
Chalk
Rocks older than Chalk

Jurassic Oxford Clay at East Tilbury

Beneath the Gault Clay at East Tilbury, Upper Jurassic Oxford Clay was found in several boreholes in a down-faulted structure extending westwards across the Thames from Cliffe in Kent. This feature is part of the complex southern faulted margin of the Anglo-Brabant Massif that was active prior to the deposition of the Lower Greensand. Its story relates more to Kent than Essex, but it is included here to illustrate that geology does not stop at county boundaries.

The sequence of strata identified in these boreholes consists of Upper Gault (about 50 m), then Lower Gault (7–8 m), then Lower Greensand (4.8 m) resting unconformably on Jurassic Oxford Clay, the base of which was not penetrated, but thicknesses up to 20 m

were drilled. Several fossil ammonites were found in the cores taken in the Oxford Clay, which proved beyond doubt the age of these strata. There is a band with phosphatic nodules at the junction between the Upper and Lower Gault which may suggest a gap in the sedimentary succession. However, there is no suggestion in the sedimentology of the Lower Gault of the proximity of land or shallowing of the sea across Essex to the north.

The ten boreholes were part of an investigation in 1959–60 into the possibility of subsurface gas storage in the thin Lower Greensand rocks, a project that was not pursued at this locality, but which shed light on this complex fault zone.

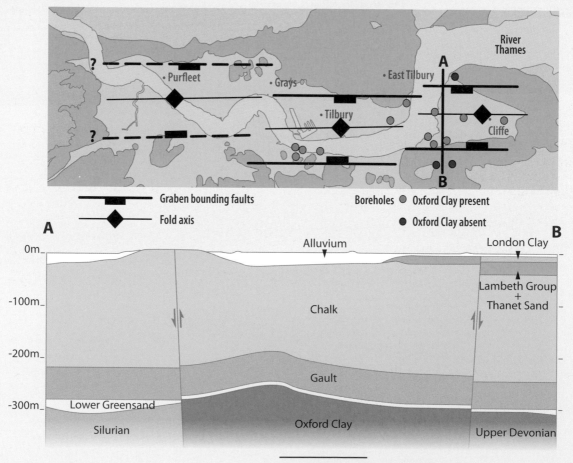

Deep movements affect the land above

The southern edge of the Anglo-Brabant Massif has acted as a thick and resistant buffer under the south border of Essex. It has been repeatedly impacted by shunts due to far-off continental shifting as Africa continues to jostle into Europe. The platform edge has been cracked by faults, which would have been the focus of multitudes of earthquakes in Essex over millions of years. Some of these fault movements have affected the build-up and structure of overlying younger rocks, influencing the landscape of Essex.

The movements produced folds and faults, such as the thrust fault discovered in the Wickham Bishops borehole drilled in 1880. Layers of rock were gently flexed as well as faulted, which may have influenced sedimentation and erosion during and after deposition. The small outcrops of older Paleogene rocks mapped near Witham indicate folding in these layers. The alignment of the Anglian ice margin with the Wickham Bishops fault along the line of the Danbury–Tiptree ridge may be more than just coincidental. Faulting at depth may explain the presence of this feature at the surface. The alignments of glacial tunnel valleys and modern river valleys seem very likely to be influenced by underlying structures, especially faults, but with the lack of boreholes and seismic surveys, there is not yet much evidence to support these ideas. However, it is extremely likely that numerous faults exist under Essex, hidden beneath and within the bedrock layers. Continued faulting, often accompanied by earthquakes, would have been frequent during the advance and retreat of ice sheets within the past 2 million years of the ice age. The crust under Essex was loaded and then unloaded as the thick ice sheets pressed down and retreated.

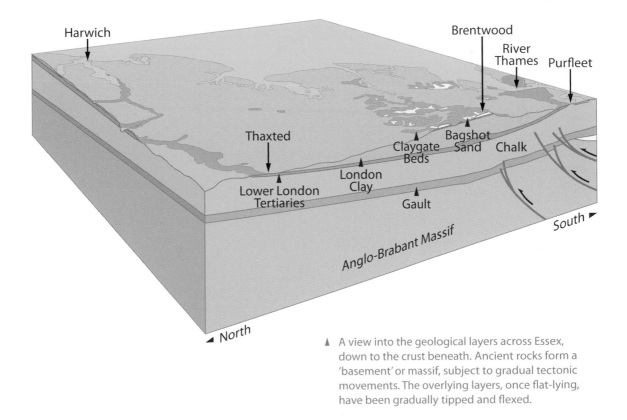

▲ A view into the geological layers across Essex, down to the crust beneath. Ancient rocks form a 'basement' or massif, subject to gradual tectonic movements. The overlying layers, once flat-lying, have been gradually tipped and flexed.

Movement along faults deep within the Anglo-Brabant Massif contributed to the Colchester earthquake of 1884, the most destructive earthquake ever recorded in Britain. The isoseismal lines constructed from records of the damage are aligned parallel to the Wickham Bishops fault. A fault some 7 kilometres (4 miles) beneath the seabed off Harwich was responsible for an earthquake in 1994, which was reported by coastguards at Walton-on-the-Naze. The relentless development of the geological structure beneath Essex continues.

Colchester earthquake 1884. ►
Illustrated Police News

▼ Tectonic features in the basement of Essex which show where movements have affected the deposition of later rock layers.
Compiled with information from Aldiss 2014 BGS Open Rept OR/14/008

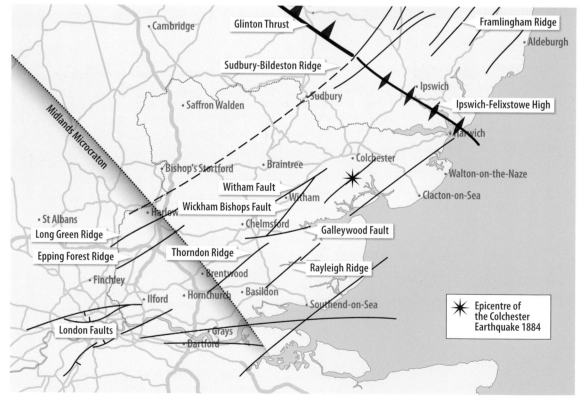

Glinton Thrust • Cambridge • Framlingham Ridge • Aldeburgh

Sudbury-Bildeston Ridge

Midlands Microcraton

• Saffron Walden • Sudbury • Ipswich

Ipswich-Felixstowe High

• Harwich

• Bishop's Stortford • Braintree • Colchester • Walton-on-the-Naze

Witham Fault • Witham • Clacton-on-Sea

Wickham Bishops Fault

• St Albans • Harlow • Chelmsford

Long Green Ridge

Galleywood Fault

Epping Forest Ridge Thorndon Ridge

• Brentwood Rayleigh Ridge

• Finchley

• Ilford • Hornchurch • Basildon

• Southend-on-Sea

London Faults • Grays

• Dartford

✳ Epicentre of the Colchester Earthquake 1884

The Purfleet 'monocline'

Southerly compression of the rocks against the Anglo-Brabant Massif has rucked up a small subsidiary fold in south Essex, the Purfleet 'monocline'. Here, the bedrock layers were affected by fault movements in the hard rocks beneath. As a result, the Chalk, which is buried beneath younger rocks across most of Essex, occurs at the surface here, a small repeat of the North Downs of Kent. The raised-up chalk layers have been exploited in many quarries around Purfleet. This uplifted ridge has also played a notable part in the story of the River Thames and the presence of humans in Essex – a story told in later chapters.

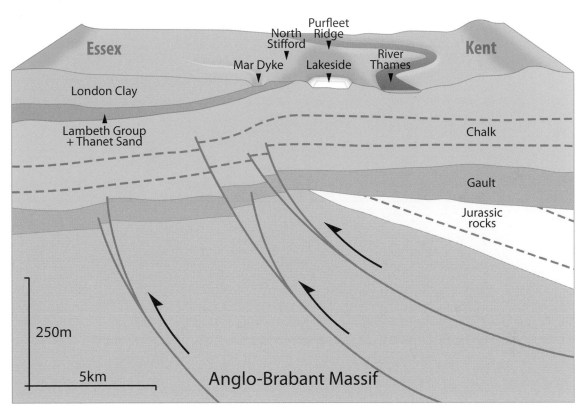

▲ Compression from the south produced the Purfleet monocline. The Chalk has been pushed upward and is now exposed within a low ridge across the Purfleet area and is seen in old chalk quarries such as Lakeside retail area. The London Clay remains in the basin to the north.

Essex has not stopped moving. Its structure continues to develop, slowly and intermittently. The edgeland of Essex continues to be subject to floods as sea level varies and the crust beneath gradually adjusts while continents drift and collide. The level of land and seabed also adjusts to the varying weight of rock layers as erosion continues, as always, to carry sediment from land to sea. Through many millions of years, the climate, seabeds and landscapes of Essex have varied dramatically. The following chapters tell the story of these changes and their effects upon today's scenery and upon our lives, starting with a great flooding of ancient lands more than 100 million years ago.

5

Chalk with bands of flint in an old quarry face. The Dartford Crossing approach road lies along the top of this cliff. Just here the rock layers have been pushed up in a low ridge by tectonic forces and erosion has cut down to the Chalk. Much of this rock was exploited for the cement industry, hence the remaining white cliffs, now mostly overgrown, where quarries were scooped out all along the ridge. Lakeside Shopping Centre occupies one such quarry.

Cretaceous	Paleogene	Neogene

145 million years   110   100   66   23   Now

Sea drowns Essex Island   ←— Chalk Sea —→   Extinction Event   Ice Age

The drowning of the island

A rise in world sea levels

It was during the Cretaceous Period that Europe, including the area of Britain, started to split away from North America. The Mid-Atlantic Ridge increased in size, causing sea levels to rise. The encroaching sea eventually overwhelmed the island of the Anglo-Brabant Massif about 110 million years ago, in mid-Cretaceous times. Early on, upland areas to the north and west were still being eroded by rivers carrying a lot of muddy sediment to the sea. When the mud settled it formed a clay-rich layer, the Upper Gault Clay. This is about 20 m (60 ft) thick under most of Essex, thickening southwards. It is composed of stiff, grey limy clay that rests directly on the inclined layers of Silurian and Devonian rocks of the eroded lands of the Anglo-Brabant Massif – Dinosaur Island. Phosphate nodules have been found in some borehole cores from the weathered top of these rocks. These were produced by chemical processes on the seabed as the sea gradually flooded across the land.

Although the Upper Gault Clay underlies all of Essex, and is seen in boreholes, it is only exposed at the surface in neighbouring counties; the fossils found there tell us that the Gault Clay was laid down in a marine environment. An exposure of Gault Clay on the Kent coast at Folkestone is renowned for its spectacular fossils, particularly of ammonites with their pearly shell layer still preserved, which have found their way into collections throughout the world. Flying reptiles called pterosaurs, with wingspans of up to 6 m (20 ft), occupied the skies over Essex during this period; the fossil evidence for this has been

▼ Pterosaurs over the sea in mid-Cretaceous times.
Jon Hughes

found in rocks of approximately the same age near Cambridge. The sea came and went, leaving some of its layers as a 'fringe' of sands around the Essex island's ever-changing coastline.

About 100 million years ago, lands to the south and west of Britain began to rise more significantly as the Atlantic was opening out. Rivers from the hilly terrain moved sand and silt into the seas around the island to make a layer known as the Upper Greensand. This is indeed green when encountered in boreholes, as it contains grains of glauconite, a green iron-potassium mineral that forms in shallow seas and estuaries. The green sandstone weathers to orange when exposed at the surface, as can be seen in Kent. From borehole evidence, Upper Greensand is known to lie beneath the southern and western parts of Essex; for instance, there is a bed of greenish-grey sandstone 4.3 m (14 ft) thick above the Gault in the deep Canvey Island borehole. Here, the sandstone contains worm tubes and ripple marks, evidence of a shallow, sandy seabed not far from the coast.

▲ A Cretaceous ammonite found in the Gault Clay at Folkestone. *Discovering Fossils website*

Sea levels continued to rise through Cretaceous times, leading to even more widespread flooding of the continents. While the last of the dinosaurs roamed across North America, almost all of Britain lay beneath a deep tropical sea in which the Chalk was deposited. A thick layer of Chalk now underlies the whole of Essex.

▼ Sandstone from Upper Greensand near Sevenoaks.

The Chalk Sea

From about 100 to 70 million years ago, and after deposition of the Upper Greensand, world sea levels continued to rise until almost all of Britain and northern Europe was submerged beneath a tropical sea. There was no ice at the poles and the Mid-Atlantic Ridge continued to swell as the new ocean was slowly forming to the south and west. The swelling caused sea level to rise by up to 300 m (1,000 ft) higher than today. The whole of Essex was under water and shorelines were far away, with the result that the sea contained very little land-derived sediment and the water was remarkably clear.

Frequent blooms of planktonic algae proliferated in the warm sea. Many of the algal forms were a type of single-celled seaweed that produced tiny calcite plates. These crystal plates grew as overlapping discs to form a spherical protective framework around each algal cell. The whole sphere is called a coccolithophore and the separate discs are coccoliths.

These microscopic algae were eaten by fish and other marine creatures such as copepod crustaceans. Their fecal pellets consisted of coccolith debris which sank to the sea bed and accumulated as layers of pure white calcite in the form of calcium carbonate mud. The white sediment gradually compacted as a unique soft white limestone called chalk.

A thickness of up to 500 m (1,600 ft) was deposited in 30 million years; the average rate of deposition is estimated to have been as little as 1 centimetre (less than half an inch) per thousand years.

▼ Submarine Essex: the geography of Britain during Late Cretaceous times, about 85 million years ago.

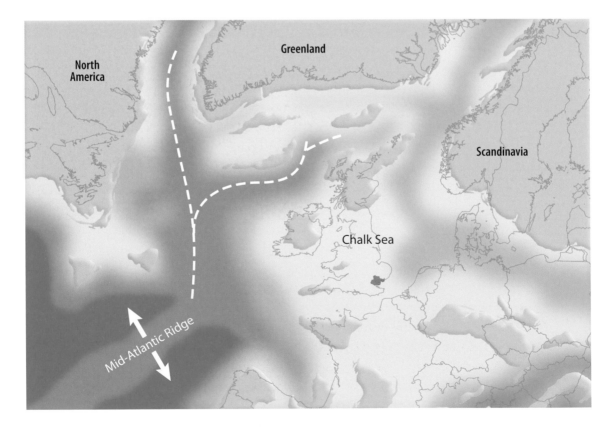

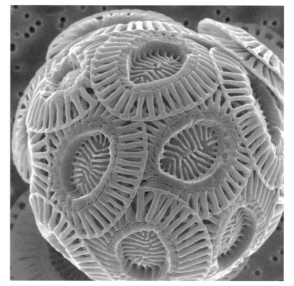

◄ A recent algal bloom
in the English Channel.
NASA Landsat image

Chalk therefore consists almost entirely of plant fossils, but they are so small that it is only possible to see them with the aid of a high-powered microscope. Even the white dust that soils your fingers as a result of examining a piece of chalk will be made up of millions of these tiny crystalline fossil remains.

A scanning electron micrograph of a single *Emiliania* ►
huxleyi cell, less than 1/100th of a millimetre across.
Wikimedia Commons Alison R. Taylor

The name 'chalk', with a small 'c', is a general name for the rock, like 'limestone' or 'shale'; however, the name of the geological unit as a whole is spelt with a capital 'C': the 'Chalk' Group, frequently referred to simply as 'the Chalk', in the same way as we may refer specifically to 'the London Clay'.

◄ A chalk hand specimen made of billions of coccolith fossils.

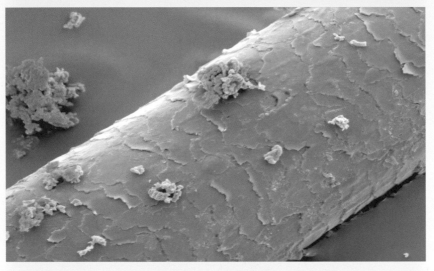

◄ Scanning electron microscope image of chalk dust on a human hair.
A single coccolith is seen in the dust particle in the upper centre.
Peter Frykman GEUS

◄ Wheel-like coccoliths clearly seen in a highly magnified image of a piece of chalk under an electron microscope. Each coccolith is only about 1/200th of a millimetre in diameter. *UNLV*

Cretaceous rocks beneath Essex

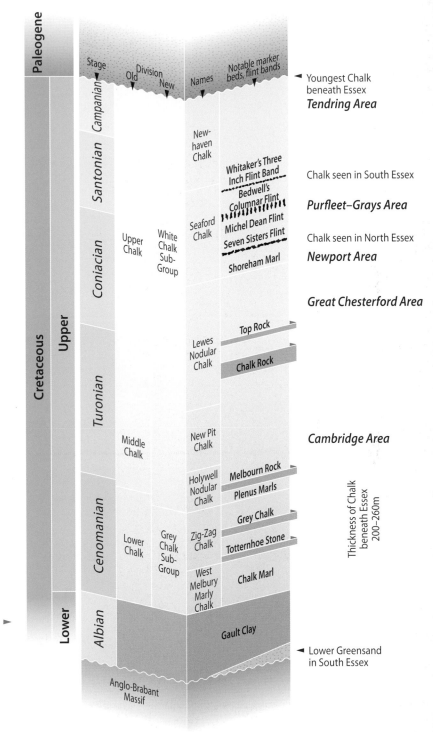

A section revealing the ages and names of the Cretaceous rocks beneath Essex. The current and previous names and divisions are given, as both are seen in accounts of the Chalk.

Stage

Division
Old New

Names

Notable marker beds, flint bands

◄ Youngest Chalk beneath Essex
Tendring Area

Chalk seen in South Essex
Purfleet–Grays Area

Chalk seen in North Essex
Newport Area

Great Chesterford Area

Cambridge Area

Thickness of Chalk beneath Essex 200–260m

◄ Lower Greensand in South Essex

Paleogene

Cretaceous

Upper

Lower

Campanian

Santonian

Coniacian

Turonian

Cenomanian

Albian

Upper Chalk

Middle Chalk

Lower Chalk

White Chalk Sub-Group

Grey Chalk Sub-Group

New-haven Chalk

Seaford Chalk

Lewes Nodular Chalk

New Pit Chalk

Holywell Nodular Chalk

Zig-Zag Chalk

West Melbury Marly Chalk

Whitaker's Three Inch Flint Band

Bedwell's Columnar Flint

Michel Dean Flint

Seven Sisters Flint

Shoreham Marl

Top Rock

Chalk Rock

Melbourn Rock

Plenus Marls

Grey Chalk

Totternhoe Stone

Chalk Marl

Gault Clay

Anglo-Brabant Massif

Naming the layers of the Chalk

Books and articles about the Chalk and its fossils may use differing names for Chalk layers, depending on when they were written. The current names of the layers in the Chalk are shown in the diagram, 'Cretaceous rocks beneath Essex' on the previous page. It also shows the older, threefold division of the Chalk: Upper, Middle and Lower, made on the basis of clay or 'marl' content, the amount of flint and the occurrence of distinctive hard bands. Zone fossils, mainly ammonites, echinoids and belemnites, together with microfossils were used to date the sequence and assist correlation.

Recent work has enabled a much more detailed subdivision based on rock characteristics – lithostratigraphy. This was particularly required for major civil engineering works such as tunnel boring for the Crossrail project and for major drainage works in East London. Geological mapping is now also based on this approach. The Chalk Group is now divided into two Subgroups, the White Chalk Subgroup and the underlying Grey Chalk Subgroup. The Grey Chalk contains more marly layers where the chalk has some clay minerals and volcanic ash mixed with it. Where these layers are distinct, they have specific names, as they can be traced between boreholes and outcrops. Recognising the subdivisions of the Chalk is important for civil engineers as the fracture properties of each formation are significantly different. The amount of flint or the number of marl layers or hard bands affect the machinery used in excavation and tunnelling.

▼ Chalk with bands of black flint in a tunnel boring.
R.N. Mortimore

When attempting to subdivide the Chalk and to correlate the divisions across the British Isles, it was found that the Chalk of Yorkshire and Lincolnshire, together with some of the beds in north Norfolk, belonged to a different depositional setting from the Chalk of Kent and Sussex and further west. However, it was acknowledged that neither sequence was wholly applicable to East Anglia and so this area is termed a 'Transitional Province'. Descriptions of the Chalk in this chapter use the most appropriate units from either scheme, and in some cases introduce locally recognised units where this is helpful, depending on the context. Much of this correlation relates to borehole data used to describe the Chalk for major engineering projects.

However, there are descriptions of the units into which the Chalk is now divided that are useful in the field as well as in boreholes. Subtle differences in colour and texture can be distinguished amid the chalky whiteness. The lowest unit in the extreme north-west of Essex, the Zig-Zag Chalk, is medium hard, pale grey and blocky and otherwise unremarkable. Above the Plenus Marl, the Holywell Nodular Chalk is creamy white with a lumpy, nodular appearance and filled with fossil shell debris. Within this unit is the Melbourn Rock, a hard layer which lacks shell material. The New Pit Chalk is blocky, greyish white and firm to moderately hard with common marl seams and very scarce flint.

The Lewes Nodular Chalk is white, creamy or yellowish and lumpy or nodular. The Chalk Rock within this unit is a very hard limestone that has been used as a building stone although, if outside, it soon shows the etching effects of acid rain. The Seaford Chalk is soft and smooth and bright white with many bands of often very large flints.

The Newhaven Chalk is soft to medium hard, smooth with numerous marl seams and flint bands.

▲ Chalk Provinces, each with differing structural features that affected sedimentation.

It includes abundant large vertical flints and distinct zones of phosphatic nodules, but is only seen in boreholes in north-east Essex as it is buried beneath younger rocks. Above this, the Culver Chalk is also a soft white chalk with many seams of generally large and sometimes tabular flints.

It is interesting to note that as the grey marly layers occur less often, the flint seams become more frequent and the flints themselves increasingly large.

The Chalk in Essex

The Essex Chalk cloaks the buried island of the Anglo-Brabant Massif. The chalk layers contain a variety of fossils that indicate a shallow sea. Beds of hard, nodular chalk occur at certain levels and these were formed in shallower seas during periods of uplift of the underlying Anglo-Brabant Massif. These movements even fractured the Chalk while it was forming. Local high areas of the seabed were caused by folds buckling over faults as they were pushed upwards, such as at Purfleet in south Essex. Due to the continuing influence of the Anglo-Brabant Massif, the Chalk in East Anglia and across to the North Downs of Kent is significantly thinner (maximum 500 m) than the sequences originally deposited in Sussex (maximum 1,500 m), Lincolnshire (800 m) and the North Sea basin (over 1,200 m).

Only the White Chalk Subgroup appears as an outcrop at the surface in Essex, although the Upper Cretaceous Chalk underlies the entire London Basin.

The lowest and therefore oldest part, the New Pit Chalk (Turonian 90 mya), is seen in the extreme north-west of the county around Great Chesterford. The Chalk is hidden beneath the overlying Palaeogene deposits south and eastwards, to reappear in the south of the county at Grays and Purfleet. The age of the Chalk beneath the erosion surface at the base of the Palaeogene becomes progressively younger eastwards. The Chalk in the top of the cliffs at Purfleet is the Seaford Chalk (Santonian 83 mya). The youngest Chalk under Essex is the upper part of the Newhaven Chalk Formation of Campanian age (75 mya) and occurs beneath the Palaeogene rock layers in the Tendring area.

▼ Map of Chalk units beneath Essex showing the age names. *Based on information supplied by Mark Woods, British Geological Survey*

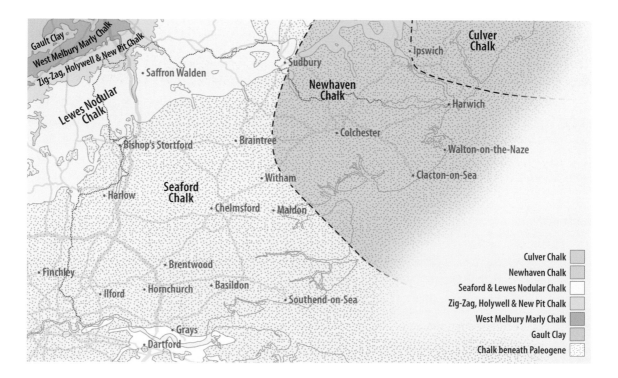

The Chalk of north-west Essex

Driving north up the M11, the rolling country north of Newport reflects the nature of the underlying Chalk bedrock. Here it is not covered by the glacial till that coats much of central Essex. The large fields take on a whitish look, especially after harvesting, and the broad valleys are usually dry. Chalk may be seen exposed in the sides of cuttings on the brows of the hills, as the road crosses progressively older layers in the Chalk succession into Cambridgeshire.

Beds of Chalk may be seen at Quendon, Newport, Audley End and Saffron Walden in the Cam valley and at Farnham and Clavering in the Stort valley. The lowermost beds in the bottom of these valleys contain several marly layers indicating deposition early in Upper Cretaceous times when muddy sediment was occasionally washed in from distant land areas. At the top of the New Pit Chalk is a hard, well-cemented band, a hardground called the 'Chalk Rock'. This is better developed towards the west in the Chilterns, but it may form the ridge of Coploe Hill on the Essex/Cambridgeshire boundary near the M11/A11 junction 9.

Above the Chalk Rock is the Lewes Nodular Chalk which grades into the Seaford Chalk. Accessible exposures are limited to old chalk pits and cuttings along less busy roads. Close examination of the vertical faces of chalk reveals details of its sedimentary structures that give clues to the fluctuating conditions on the bed of the Chalk Sea. Marl seams often only a few centimetres thick result from an increased supply of land-based material, mainly clay minerals from mud together with minerals of volcanic origin perhaps associated with the opening of the Atlantic Ocean.

▼ Construction of the M11 through the Chalk in north-west Essex in the late 1970s. The pure white rock is still to be glimpsed in some of the motorway cuttings. *Essex Field Club*

These beds are most common in the lower part of the Chalk succession and we only see the last remnants of these at the surface in north Essex. Nodular fabrics represent periods of reduced sedimentation rates when the chalk ooze between burrows in the seabed became cemented. Subsequent winnowing by currents washed away the softer ooze and the nodules were rolled around on the seabed before they became trapped by sediment again. During prolonged breaks in sedimentation the whole layer on the seabed became cemented and then encrusted and bored by various organisms. Sometimes there is also mineralisation by iron compounds – pyrite, limonite, glauconite and phosphate – that stain the otherwise pure white chalk in a variety of colours. These layers are called 'hardgrounds' and provide useful marker horizons for correlation, but their greater resistance also creates difficulties when it comes to engineering.

About 6 m (20 ft) of the Seaford Chalk is exposed in the cliff face at Limefields Pit, an old chalk quarry in Saffron Walden in which new housing has been built, but part of the quarry face is still visible. A total of about 20 metres' thickness of chalk was visible in the quarry in 1932 although this was not all in one face. During construction of the houses on the floor of the quarry a considerable quantity of soil was imported and the lower part of the chalk face is now buried including a marl seam marker bed. The chalk is soft, white and blocky, with widely spaced courses of nodular flint in layers and oblique veins of tabular flint. Fossils are now surprisingly rare, but were found in particular layers when the pits were worked by hand, enabling the dating to be determined as

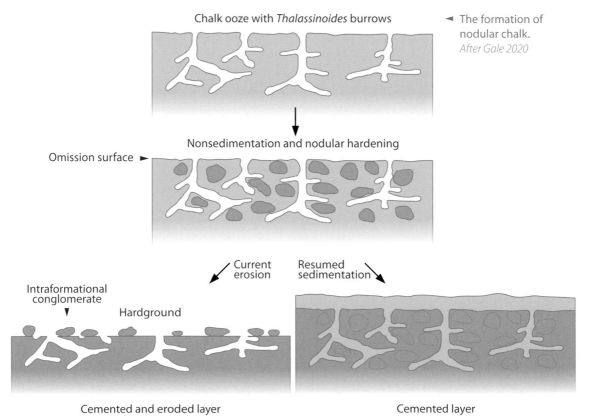

Chalk ooze with *Thalassinoides* burrows

◄ The formation of nodular chalk.
After Gale 2020

Nonsedimentation and nodular hardening

Omission surface ►

Current erosion

Resumed sedimentation

Intraformational conglomerate

Hardground

Cemented and eroded layer

Cemented layer

of Coniacian age (88 mya), in the lower part of the *Micraster coranguinum* (a type of echinoid) zone. This has enabled the Chalk here to be fitted into the bigger picture showing the way the beds change in thickness laterally and with time. Much scientific work remains to be done to relate all the details of the Chalk in East Anglia, including Essex, into the regimes demonstrated in the Northern and Southern Provinces. Most of the boreholes that penetrate the Chalk were drilled as water wells up until the 1950s, before modern methods of borehole logging using wireline tools were developed; this has hampered scientific research.

Chalk in Limefields Pit, Saffron Walden with close-up ▼ showing evidence of burrowing by marine creatures in the Chalk Sea floor.

There is no doubt that chalk has been used from time immemorial for 'chalking' the land to improve the soil. Today (2022), Newport Quarry is the only working chalk quarry in Essex, producing ground chalk and lump chalk for agriculture and industry.

The pit was first referred to by Whitaker in 1878 as a chalk pit with flints east of Newport Station containing 'a very persistent horizontal layer of tabular flint thrown down 18 inches by a fault, and two large pipes of gravel stop at this layer'. Whitaker also lists fossils found in this pit – fragments of *Inoceramus* bivalve shell and the sponge *Ventriculites*. The quarry provides an exposure of 20 metres of fractured White Chalk with a persistent layer of tabular flint that occurs about 6 metres below the top of the face. Fossils are, unfortunately, now very rare and no more precise dating is given by the British Geological Survey. Fitting the layers found here into the lithostratigraphic scheme that uses marker beds and the nature of the chalk itself, will provide significant scientific detail. For this reason alone, it is important that vertical faces in former chalk quarries are planned and conserved in restoration schemes.

Open, rolling tracts characterise the farmlands of north-west Essex where the Chalk is at the surface, for example near Saffron Walden and Great Chesterford,

◄ Chalk scenery near Great Chesterford close to the north-west Essex border. The trees across the valley are growing in a patch of glacial till.

▲ Newport Chalk Quarry in north-west Essex. *Dominic Davey*

▲ **View of Newport Quarry.** *Google Earth*

and this countryside is easily recognised when travelling along the M11. The soil on the uplands bordering the Cam valley and extending east towards Chrishall in Essex, the highest point of the East Anglian Heights at 147 m (482 ft), is dry and chalky, the white chalk showing in the fields particularly after ploughing.

The Chalk is one of the most important underground water sources (aquifers) in England. In the past, considerable quantities of groundwater were abstracted to meet demand, in particular for potable and irrigation water supplies. In north Essex and west Essex some of the water supply still comes from boreholes into the Chalk. This rock has excellent aquifer properties, having plenty of pore space between the microscopic particles and within small fractures. Crucially, it has good permeability; the interconnected network of fractures allows water to flow or percolate through the solid mass of rock.

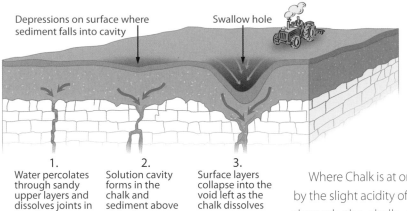

Depressions on surface where sediment falls into cavity

Swallow hole

◄ Swallow hole formation in the Chalk below a covering of permeable sand.

1.
Water percolates through sandy upper layers and dissolves joints in the chalk

2.
Solution cavity forms in the chalk and sediment above sinks down

3.
Surface layers collapse into the void left as the chalk dissolves

Where Chalk is at or near the surface, it is dissolved by the slight acidity of rainwater, which erodes down through the chalk, often exploiting pre-existing fissures and fractures. As a dissolved cavity enlarges, solution gaps form and the overlying sediment and soil collapses to create a hole or depression at the surface. Water flowing along streams may suddenly disappear down these features, hence their name 'swallow holes'. Horses and ploughs and even tractors have been known to fall into such holes in the fields of north Essex where the Chalk lies just beneath a thin surface covering of younger rocks and soil.

▼ Swallow hole, Hill Farm, Gestingthorpe, north Essex shortly after surface collapse in 2002. *Ashley Cooper*

The white cliffs of Purfleet

If you approached London up the Thames from Tilbury in the eighteenth century, a prominent feature of the landscape would have been the white cliffs of Purfleet as shown in the engraving from 1807. The river cut into the side of the Purfleet monocline and exposed some 30 m (100 ft) of the Chalk in the lofty Beacon Cliff.

These spectacular cliffs had long been exploited for lime. *White's Directory* of 1848 states: 'Near it (Purfleet) are the extensive lime and chalk pits of W.H. Whitbread, Esq., the lord of the manor. Many thousand tons of lime are burnt annually, and sent to London and other places; and from the kilns, railways are extended to the quarries, as well to the shipping. The chalk cliffs appear to have been worked for several centuries.' Deneholes, thought to be medieval chalk mines, are found nearby in Hangman's Wood, Little Thurrock.

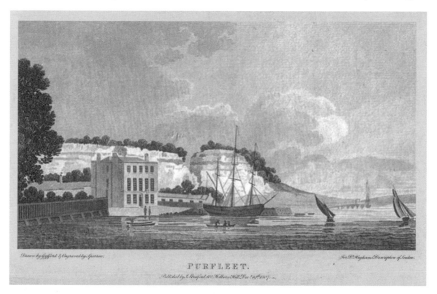

◄ An engraving showing the white cliffs of Purfleet with the Royal Hotel in the foreground. *© National Maritime Museum, Greenwich, England*

◄ Royal Hotel, Purfleet today. *Royal Hotel*

◄ The Grays Chalk and Whiting Company Quarry at Grays, Essex. They discovered water at the bottom of the pit and built a pumping station in 1863 to pump water to London.
Thurrock Museum

▼ Chalk cliff in south Essex: Chafford Gorges near Grays.

▲ Remnants of The Rock at the entrance
to Botany Quarry, Purfleet, 1930.
It was once over 30 m (100 ft) high.
Purfleet Community Forum website

▲ Beacon Cliff, Purfleet with its experimental
lighthouse. In August 1828 Trinity House decided
to build a lighthouse for the purpose of testing
the various types of lamps and lamp oils, reflectors
and lenses that were being invented or improved
upon by the new generation of inventors.
Thurrock Museum

Final demolition of the Rock in 1933. ►
Press cutting from The Sphere, British Newspaper Archive

THE CHALK CLIFF, NEAR PURFLEET, from which Queen Elizabeth
is said to have watched her fleet in the Thames, is being cut away
to make room for new buildings.   This interesting geological
formation is a result of a fold in the strata, which brings the
chalk above the level of the rest of the river bank. Being hard
the chalk has resisted erosion to an unusual extent

Today, there are many giant quarries between Purfleet and Grays, the remnants of the Portland cement industry that began in the middle of the nineteenth century. Lakeside Shopping Centre is situated in one of these quarries. In 1874 the Tunnel Portland Cement Company started quarrying chalk to the east of what became the Dartford Tunnel approach road (the A282 Canterbury Way). The Tunnel Cement Works Quarry, also known as the Motherwell Way Quarry, provided chalk to the giant Tunnel Cement Works and became one of the largest chalk quarries in Britain. From 1927 a pipeline carried liquid London Clay from the clay pits at Aveley to be mixed with the chalk to make cement. By 1968 the Tunnel Cement Works was the largest in Western Europe with 1,200 employees. Production ceased in 1976.

Except for a small fragment of Beacon Cliff, the white cliffs of Purfleet are no more. Queen Elizabeth I is said to have viewed her fleet from the clifftop, though there is no historical record of this.

The Chalk in south Essex is younger than that found in north-west Essex and is more similar in aspect to the North Downs of Kent. This area may be thought of as a 'little bit of Kent in Essex'. It is Seaford Chalk of Santonian age (83 million years) and contains several prominent bands that can be correlated over long distances.

▼ Seaford Chalk in cliffs of the old Dolphin Quarry at Chafford near Grays, with Whitaker's Three Inch Flint Band labelled.

Whitaker's Three Inch Flint Band

As engineering projects, particularly tunnels, go deeper beneath East London and Thurrock, the more frequently they are constructed in the Chalk. A ground model has been developed using boreholes and outcrop information from the old quarry faces and new developments such as the M25 Thurrock services, Lakeside and Bluewater shopping centres and the Channel Tunnel Rail Link. Important aspects include the size and frequency of flints, flint bands and hardgrounds that affect ground resistance, permeability and groundwater flow. Eleven marker beds, surfaces or broad horizons have been identified and a summary of the engineering properties is shown in the diagram. For example, Whitaker's Three Inch Flint Band is a marker, easily distinguished in the cliff of the old Dolphin Quarry (pictured facing page), that consists of a thin sheet of connected nodular flints which can be traced across Essex, Kent and into France.

▼ Composite section through Chalk of East London and Thurrock with marker bands and engineering descriptions. *After Mortimore et al. 2011*

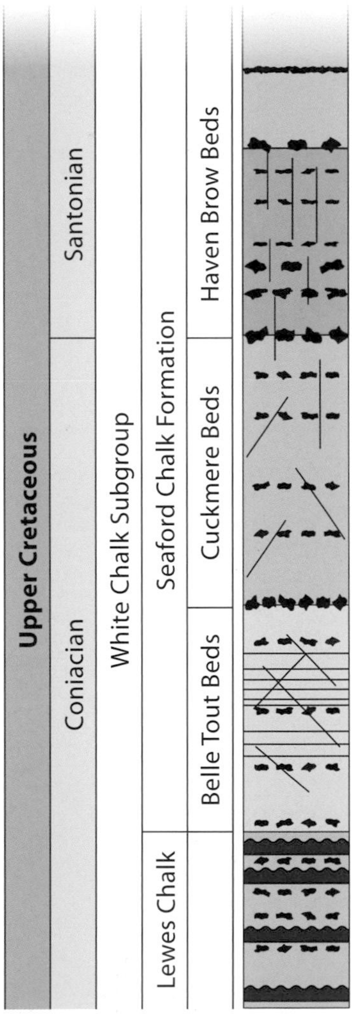

					Marker	Description
Upper Cretaceous	Santonian	White Chalk Subgroup	Seaford Chalk Formation	Haven Brow Beds	**Whitaker's 3" Flint Band** / **Bedwell's Columnar Flint**	Extremely weak to very weak, low and medium density, firm chalk with regular flint bands predominantly sub-vertical, closely spaced (<50mm), clean closed joint sets (heavily stained and mineralised where fracture zones and groundwater flow are encountered) otherwise closed and clean
	Coniacian			Cuckmere Beds	**Michel Dean Flint**	Extremely weak to very weak, low density, chalk (the beds with the most drilling damage) with regular flint bands sub-vertical, closely spaced, clean closed joint sets (heavily stained and mineralised where fracture zones and groundwater flow are encountered) becoming widely spaced; steeply inclined 70° dipping joints at the base
				Belle Tout Beds	**Seven Sisters Flint Band** / **Belle Tout Marls**	Very weak to weak, firm shelly chalk low and medium density, with regular flint bands and thin marls sub-vertical, widely spaced closed, heavily slickensided (polished and striated) steeply inclined 70° dipping joints
			Lewes Chalk		**Shoreham Marl 2**	Alternating very weak to moderately strong and very strong firm, coarse to rough nodular chalk and hardgrounds low, medium to high density chalk with regular flint bands poorly to irregularly fractured

Life in the Chalk Sea

The sea in Upper Cretaceous times was home to many creatures, some of whose fossilised remains can be found embedded in chalk and flints. The most common of these fossils found in the Chalk of Essex are sea urchins (echinoids). Some were bottom-dwelling scavengers such as *Cidaris* and others, such as *Micraster*, lived in burrows in the seabed.

The chalk itself looks fine-grained and white to the casual observer. The individual grains of coccoliths, producing the fine powder when rubbed with a finger, are much too small to see.

The Chalk Sea floor: a reconstruction showing ►
swimming animals – sharks, scallops, ammonites and belemnites – with sponges, sea urchins (echinoids), sea lilies (crinoids), mussels and other bivalves living on the white mud of the seabed. Echinoids, lobsters and soft-bodied worms burrowed into the seabed. From a diorama in the former Geological Museum, London. *Alamy*

Along with these minute calcite crystals are micro-fossils that can only be seen with a high-powered microscope; some have beautiful coiled shells of white calcite. These tiny fossils are very useful for age-dating, especially from boreholes where the chance of coring a recognisable echinoid or bivalve shell is very small.

Microfossils ►
separated from a chalk sample. These are identified to provide relative dating and correlation of beds in the Chalk succession.

1 mm

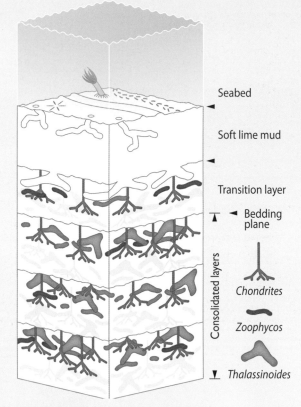

Seabed

Soft lime mud

Transition layer

Bedding plane

Consolidated layers

Chondrites

Zoophycos

Thalassinoides

Trace fossils

There is little evidence of small-scale bedding in the Chalk. It looks very uniform between the obvious bands of flint. This is because the sediment has been thoroughly disturbed – bioturbated – after passing through the guts of various creatures that lived within the seabed. Staining chalk with a coloured liquid can show up the shapes of burrows that were made by lobsters and shrimps and other animals such as worms. These trace-fossil markings are very like the structures of burrows within modern beaches and seabeds. They have been given names as if they were fossils of creatures, but, although the shapes are quite distinctive, there is very little evidence of the actual animals that made them.

◄ Trace fossils in the modern seabed.
After Savrda 2007

▼ Burrows in the Chalk stained and recoloured. *Andy Gale/PGA*

▼ Trace fossil forms in the Chalk.

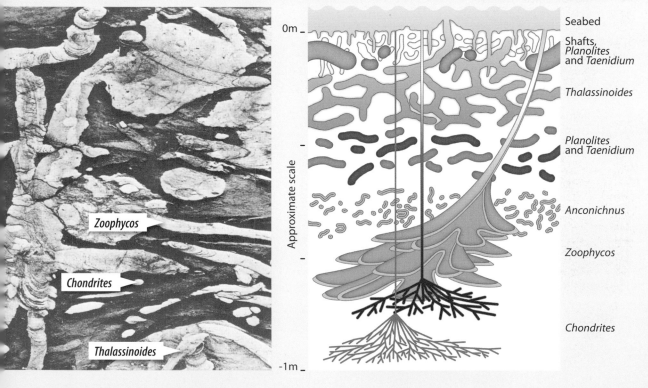

Zoophycos

Chondrites

Thalassinoides

0m

Approximate scale

-1m

Seabed

Shafts, *Planolites* and *Taenidium*

Thalassinoides

Planolites and *Taenidium*

Anconichnus

Zoophycos

Chondrites

Macro fossils

The only fossils to be commonly preserved in the Upper Chalk are those with hard parts such as bivalves (a class of mollusc which includes cockles, mussels, clams and oysters), brachiopods (another type of shellfish), crinoids (sea lilies), sponges and corals. Teeth of various kinds of fish, including sharks, are also found.

The shells of gastropods (a class of mollusc which includes snails, whelks and limpets), ammonites (extinct molluscs with coiled, chambered shells) and some species of bivalve were made of a form of calcium carbonate, aragonite, that subsequently dissolved in the chemical environment of the Chalk seabed and so are not usually preserved as fossils.

▲ Fossil echinoid *Cidaris*, from Botany Quarry, Purfleet, complete with spines that protected the animal during its life. *Gerald Lucy; specimen courtesy of Michael Daniels*

◄ Starfish *Crateraster quinqueloba* fossil preserved in flint. *David Turner*

◄ *Marsupites testudinarius. UKRI NERC*

▲ Fossil echinoid *Micraster* from a chalk quarry near Grays. *UK Fossils files*

◄ *Uintacrinus socialis. Wikimedia Commons*

▲ Belemnite in chalk

Special adaptations

Many of the fossilised animals show special adaptation to conditions on the bed of the Chalk Sea. The ooze of coccoliths and other microfossils was a very soft, soupy mix so animals developed shells with long spines or extra-large surface areas or bodies with long tethers to enable them to be almost suspended on top of the white mud. The large oyster-related clam, *Inoceramus* had a huge flat shell, in some species over a metre across, to act like a snowshoe. These shells were quite thick and made of prisms of calcite that often fell apart as the shell was incorporated into the lime mud, giving rise to rectangular-shaped fragments, often seen in flint as well as in the chalk itself. There is a remarkable record from 1870 of the under-surface of a bed of chalk covered with the impressions of magnificent entire shells of *Inoceramus*. This was found when the tunnel from Grays Chalk Quarry to Titan Quarry was dug under Hogg Lane. Meeson, the quarry owner, had the surface cleared and a number of casts were taken.

▲ The fossil oyster *Inoceramus* preserved in flint from Saffron Walden. The large 'wing' on the shell acts like a snowshoe. The hinge is about 7 cm long.
Gerald Lucy

1 cm

▲ The fossil bivalve *Spondylus spinosa* from the Chalk. The spines supported the shell on the soft seabed.
© *Museums Victoria*

▲ Modern shell of *Spondylus spinosa*.
Wikimedia Commons/Didier Descouens

Finding Chalk fossils

In Essex, fossils occur in virtually any Chalk exposure, but spotting them requires some experience, as they are often nearly the same colour as the host rock. There are large disused quarries near Grays and Purfleet, which produced some magnificent fossils during their working life, but access is restricted nowadays and permission must always be sought before entering. Many of these old quarries are flooded, with high vertical faces. Some have been infilled with waste and become nature reserves. Unfortunately, the geological value of these sites is not always appreciated and many of the faces are threatened by landscaping or excess wild plant growth.

In the north of the county several small, old pits in the Upper Chalk near Saffron Walden have also yielded specimens, and many Chalk fossils were found during excavation of cuttings for the M11 motorway in the region. Waterworn flint pebbles, derived from the Upper Chalk, can be found throughout Essex and these often contain fragments or traces of fossils.

▼ Fossil echinoid (sea urchin) found in Chalk during construction of the M25 motorway in Thurrock. *Essex Field Club*

Fossils from the Chalk & Flint

Sea Urchins

Cast of cidarid *Tylocidaris*
with spines

Cidarid
sea urchin spines
are often found separately

Part of a cidarid
preserved in a flint nodule

Micraster
view of base preserved in chalk

Micraster
Top view, preserved in flint

Echinocorys
preserved in chalk

Conulus
preserved in flint

Bivalves

Spondylus spinosus
spines (broken off in this specimen)
help it to rest on the surface of the chalk sediment

Fragment of
Inoceramid,
Birostrina sulcata
preserved in flint

Fragment of an
inoceramid shell
preserved in flint

Inoceramid, *Birostrina concentrica*.
Large size helps to prevent sinking into the chalk sea bed

Internal mould of inoceramid, *Volviceramus*
hinge area preserved in flint

Belemnite

Brachiopods

Rhynchonelid,
Cyclothyris latissima
preserved in chalk

Mould of
rhynchonelid
in flint

Terebratulid, *Gibbithyris semiglobosa*
preserved in chalk

Fish teeth

*Special specimens
collected from
Essex Chalk pits
when they were
worked by
hand*

Teeth from a shell-crushing shark *Ptychodus polygyrus*
in a flint nodule collected in 1915
from Grays, Essex

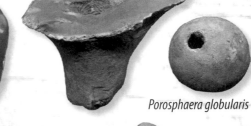

Shark teeth *Cretoxyrhina mantelli*
preserved in chalk, collected in 1850
from Grays, Essex

Sponges

*These are often
preserved in flint,
but it is seldom
possible to
identify particular
species*

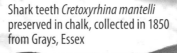

Porosphaera globularis

Chalk Sea monsters

At the top of the food chain in the Chalk Sea were mosasaurs, giant marine reptiles which were up to 10 m (33 ft) long and must have been formidable predators. A mosasaur had a long body and tail, paddle-like limbs and heavy jaws armed with sharp, conical teeth. Most of the mosasaur remains from the English Chalk are fragmentary; the fossils from Essex include isolated teeth found by quarry workers at Grays in the nineteenth century. Part of a mosasaur jaw, with 5 cm-long (2 inch) teeth filled with flint, an impressive specimen, was found by palaeontologist Edward Charlesworth at Grays in 1835 (now in the Natural History Museum, London).

▼ Engraving of a mosasaur jaw found in the Chalk at Grays in 1835.

A reconstruction ▲ of a mosasaur. *Xactimate*

Mosasaur skeleton ► in Maastricht Natural History Museum in the Netherlands. *Wikimedia Commons/ Wilson 44691*

Mosasaurus stenodon, from the Essex chalk.
Flint Nucleus occupying the Pulp Cavity of the Tooth.

W.Bowman, Del.et Lith.

Printed by W.Monkhouse, York.

The origin of flint

Layers and nodules seen in chalk cliffs, and numerous pebbles on beaches and in gravel, are made of flint. Flint is silica (silicon dioxide) in a very finely crystalline state. It is a hard, tough substance, in great contrast to the soft, powdery chalk in which it occurs. Flint is seen in a bewildering variety of colours, shapes and forms and its origin has long been the subject of debate, but a consensus relating to changing chemical conditions within the Chalk seabed is now emerging.

The silica of flint was derived mostly from the skeletons of glass sponges growing on the seabed. The tiny skeletal parts, a scaffolding of spicules within the sponges, were dispersed into burrows in the chalk mud whenever the sponges disintegrated. These spicules were subsequently dissolved by alkaline seawater trapped in the limy mud; the silica was precipitated initially as silica gel and this later crystallised as minute quartz rods – silica crystals – to form the mass of flint.

This sequence occurred at certain intervals in the seabed, mainly in burrows made by marine animals, as these were more permeable, to form well-defined, parallel bands of flint nodules at many levels within the Chalk. These can be correlated across wide areas, each with its own characteristic type of nodule and associated shell fossils. An example is Whitaker's Three Inch Flint Band that can be seen near the top of the cliff at Chafford. This flint band is a sheet-like network

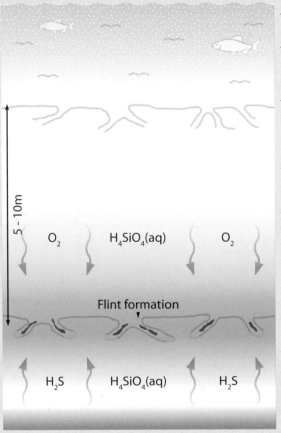

▲ Stages in the formation of flint within and around animal burrows in the seabed.

◄ Siliceous plankton – radiolaria, diatoms.

◄ Alkaline sea water dissolves silica.

◄ Glass sponges – spicules made of silica. Burrowed sea floor made of lime mud.

◄ Decaying plants and animals in lime mud of sea bed use up oxygen.

◄ Organic silica (biogenic opal) is dissolved.

Zone of mixing

◄ Hydrogen sulphide + oxygen = sulphuric acid. Hydrogen + calcium carbonate = calcium + carbonic acid ions. Carbonic acid ions help precipitate silica – *flint*. High porosity in burrows leads to their preferential use for chemical reactions that form flint.

◄ Decay of organic material without oxygen produces hydrogen sulphide. The low iron content of chalk mud precludes the formation of iron sulphide that would take place in iron-rich muddy sediments.

A flint revealing the shape of the burrows in which it formed.
Gerald Lucy

of burrow infills that can be traced across north Kent and right across the Channel into the Paris Basin.

When freshly dug from the Chalk, flint nodules are commonly black or dark grey in colour, with a white outer rind or 'cortex'. Flint nodules can be of the most remarkable shapes, often with spiky protrusions; sometimes they can be ring-shaped or inter-locking structures. These shapes largely result from the shapes of the burrows within which the flint formed. Such flint shapes are often mistaken for the fossils of all sorts of bizarre creatures, or of teeth or antlers. Banded flints also look like fossils; the banding occurs when solutions of a slightly different chemistry flow through flint that is still a gel. The bands often weather differentially giving a corrugated or organic, fossil-like appearance.

Banded flints: although these look remarkably like fossils, the patterns were created by inorganic chemical diffusion bands through the flint as it was forming.

Features of flint

Many fossils in the Chalk, particularly echinoids, are found to be infilled with or enclosed by flint. Echinoids are of particular interest because when the shell is removed using dilute acid, the flint surface beneath often reveals the very finest details of the underside of the shell.

Flints often contain fossil sponges which are usually preserved as irregular hollow cavities in flint nodules; when broken open these nodules sometimes contain a white powder known as 'flint flour' which contains sponge spicules and microscopic fossils of tiny animals that lived on and in the sponge during its life.

Occasionally, cavities lined with sparkling crystals of transparent quartz or smooth masses of chalcedony are found in flint, formed by recrystallisation of silica, but these are quite rare.

Flint is very tough, but it fractures to produce razor-sharp fragments with a characteristic 'conchoidal' fracture surface, so called because it resembles the surface of a shell. With such properties, flint has long been vitally important throughout human development for the manufacture of tools and for making fire.

▼ Part of an *Inoceramus* bivalve shell preserved in flint in a church wall at Willingale.

Fossil echinoid, ►
Conulus, made of
flint. This is a flint cast
that preserves the
detail of the inside of
the original shell.

Flint showing ►
its characteristic shell-like
or conchoidal fracture.

▲ Flint with quartz
crystals lining
a cavity.

▲ A pebble of flint
that has been cut
in half to reveal
a fossil sponge.

▲ A Palaeolithic flint culinary tool: a serrated ficron
handaxe used to prepare mammoth steaks.

Mapping the age of the Chalk across Essex shows that the age of the topmost Chalk gets younger towards the north-east of the county. Mapping also shows that there is a lot of Chalk missing between the Chalk and the overlying Thanet Sands. The amount of missing Chalk increases towards the west where only the beds at the bottom of the White Chalk Group are present beneath the Thanet Sands. We know that the higher, younger beds were once deposited as they occur under Norfolk and the North Sea.

During and after the deposition of the Chalk, there was increased volcanic activity above the Iceland mantle plume as the Atlantic continued to rift open. This tectonic activity tilted the crust around the new rift. The thick layers of Chalk were raised above sea level across Britain from Ireland to the North Sea, tilting them from the north-west. The Chalk was rapidly eroded, shedding enormous numbers of flints onto the landscape. In west Essex, erosion penetrated down to the middle of the Lewes Nodular Chalk; estimates suggest that some 500 m (1,600 ft) of Chalk was removed by erosion in 10 million years. In eastern Essex about 350 m (1,100 ft) were eroded. Fractures in the exposed Chalk were widened into fissures, lined with flints and then filled by sand as we shall see in the next chapter. The released flints were ground down and rounded on beaches to be redistributed across Britain and incorporated into the next major layers, preserving the remaining Chalk beneath.

Flint pebbles

The Chalk across and to the west of Essex was eroded very rapidly as the land started to tilt. Billions of flints were released as the soft rock crumbled and dissolved.

The flints were carried by rivers and deposited on beaches as seas came and went. Wherever there was a seashore, the waves rolled the flints until they became beautifully rounded. Storms pounded the cobbles and pebbles, causing internal fractures that show up on their surfaces as small, curved 'chatter marks'. Rapid erosion continues to this day in the cliffs of southern England, most spectacularly where the sea is eating into the chalk downs of Kent and Sussex causing many cliff falls and forming extensive flint beaches.

Fresh out of the chalk, flints often have a white coating or 'cortex', due to interaction with the chalk during its formation. This is gradually worn away during transport. The interior of fresh flint appears black as it is made of tiny transparent crystals – the light goes in but cannot escape. Depending on their history, flints take on many colours. Flint is microporous and solutions, often containing iron, seep into the flint, colouring it brown or orange.

Chatter-marked flint

Coloured flint

Cortex-covered flint

Black flint

Manganese in sea water gives flint a black coating, so you see black flint pebbles with an orange interior. Within gravel deposits, flint surfaces may be partly dissolved to give pale grey or white coatings. Some flints are marbled or mottled where part of the original interior and part of the outer coating are both at the surface. Flints heated in forest fires turn salmon pink or red and are sometimes crackled and white.

Curious shapes result where the original shapes of flint burrow fillings have worn into smooth, elongated or branching nodules looking like 'bones', 'teeth', small 'animals' and other fanciful forms. Together with holes, fossils, banding and other markings, the long and eventful histories of countless flint pebbles have created an intriguing variety across the region. Essex gravel is seen in footpaths, seashores and many ancient church walls. Their content provides a thought-provoking story for geological insight and artistic inspiration.

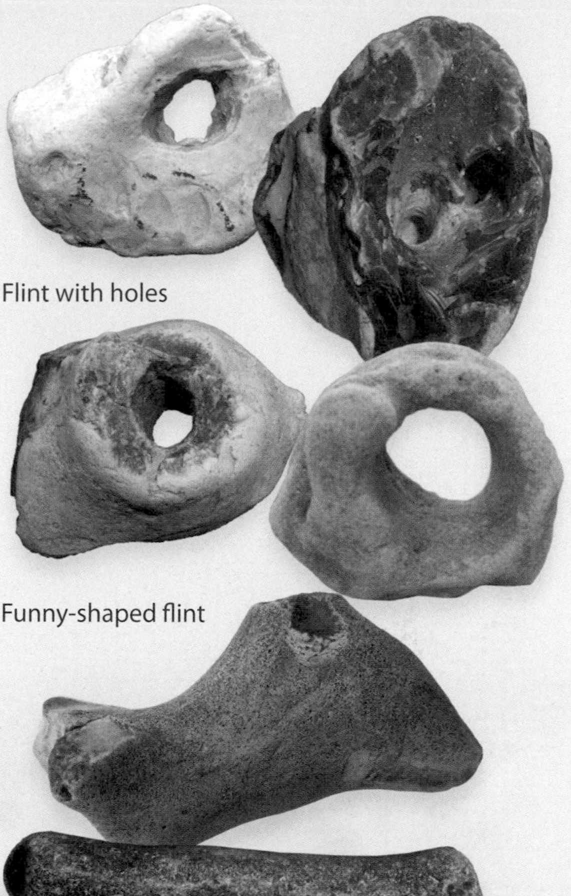

Flint with holes

Funny-shaped flint

Mottled flint

Banded flint

Heated flint

Black coated flint

Chalcedony in flint

6

A seabed full of bivalve shells was discovered in the excavations for the new M25 motorway in 1979. Known as the Oldhaven Beds, this sandstone layer is evidence of a shallow sea across Essex 55 million years ago.
Essex Field Club

Paleogene   Neogene

66 million years   55.5   23   2.58 ▲   now
      ▲   Edgeland Essex   ▲
Extinction Event

Seashores and swamps

Paleogene rocks beneath Essex

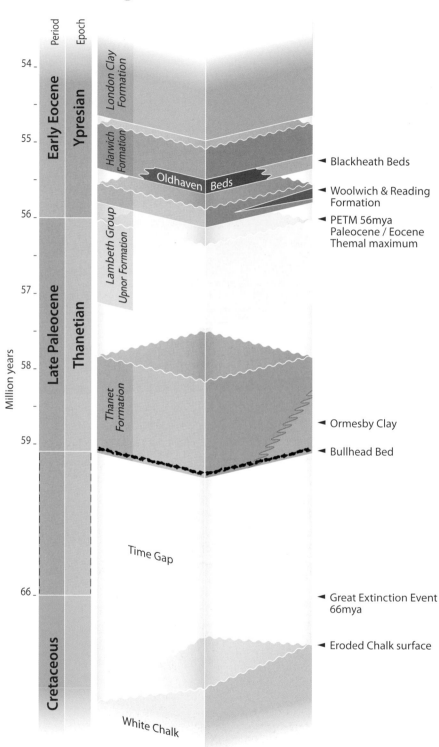

A great extinction

Around 65 to 60 million years ago Britain and Ireland were uplifted from the west as the North Atlantic Ocean was beginning to form. With all the volcanic activity at the new mid-ocean ridge, molten rock – magma – moved within the depths beneath the ocean floor, tilting the crust beneath Britain upwards from the west. This uplift caused rivers to flow eastward across the area of southern Britain. The ancestral Thames flowing from north Wales to East Anglia may have been initiated at this time. So, the Chalk Sea retreated eastwards and the new land surface, with its thick chalk layer, was very rapidly eroded. The North Sea region, where rifting had started in Jurassic times, has continued to subside with thick layers of sediment being laid down right up to the present day.

The junction between the Chalk and the overlying Paleogene sediments represents a time gap during which a mass extinction took place throughout the world 66 million years ago – the Cretaceous-Paleogene 'K-Pg' boundary (previously called the Cretaceous-Tertiary or K-T boundary). The actual time of the extinction is not recorded in Essex, as the topmost beds of the Chalk have been eroded and are missing here. The lowest part of the overlying Thanet Formation is dated at around 59 million years. Nevertheless, this junction in Essex still represents one of the most important events in the history of the Earth, a time that spelled extinction not only for dinosaurs and the giant marine reptiles but also for millions of other creatures including the ammonites. The impact of a giant meteorite is at least part of the explanation.

This extinction event ended the Mesozoic Era. The sea has returned to the area of Essex many times since then and the countless millions of flint nodules from the eroded horizons of the Chalk have been ground down to cobbles, pebbles and sand by the relentless pounding of the waves. These worn-down and often very rounded flints have been redeposited within many of the rock layers described in the following chapters.

Thanet Sand fills ▶
solution hollows in
the Chalk surface at
Chafford Gorges near
Grays in south Essex.

Edgeland Essex

The rock layers that overlie the Chalk are a complicated sequence of sands, silts, clays, pebble beds and shell beds that vary across the county. These were laid down at the margins of the southern arm of a shallow sea that came and went across the area of the present London Basin, southern North Sea and into Europe over a period of some 4 million years. These beds are tricky to date as they do not contain diagnostic, rapidly evolving fossils, so correlation has been difficult. The different rock names used in the literature can be confusing. In the early geological mapping by Sir Joseph Prestwich and others in the nineteenth century, the rocks between the Chalk and the London Clay

were referred to as the 'Lower London Tertiaries' (LLT). We use this term where it is more convenient to refer to this whole sequence of beds. The term survives on some published geological maps covering northern Essex, where it refers to 'undifferentiated Harwich Formation, Lambeth Group and Thanet Formation'. As we now understand more about the correlation of this complex sequence of beds and the changing environments of deposition, more detailed terms are used to describe the story in this chapter.

The comings and goings of the sea are shown by the changes between successive layers. Identification of these varied rock layers is based largely on the correlation of the different types of sediment

◄ After the Great Extinction the sea washed back and forth across the Essex edgelands. Fine sand and seams of rounded flint pebbles tell the story at Orsett Depot Quarry, West Pit.
British Geological Survey © UKRI 2021

▼ The Lower London Tertiaries (LLT) underlie much of Essex. The surface outcrop appears as a thin band of sandy rocks around the edge of the overlying London Clay.

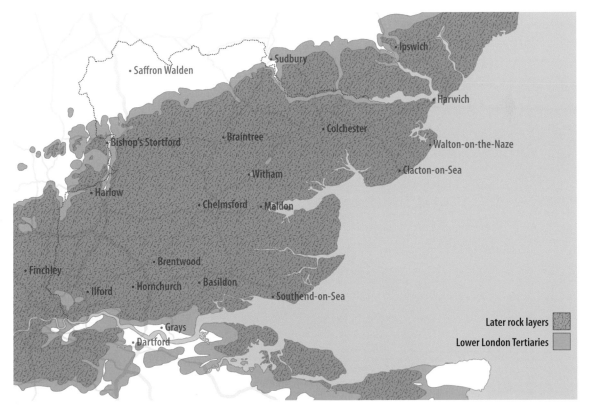

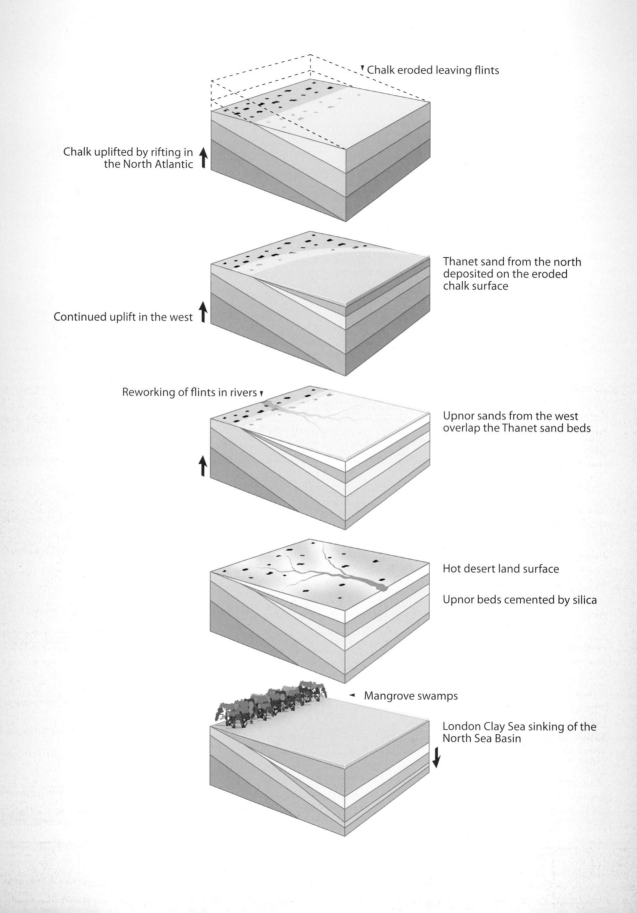

Chalk eroded leaving flints

Chalk uplifted by rifting in
the North Atlantic

Thanet sand from the north
deposited on the eroded
chalk surface

Continued uplift in the west

Reworking of flints in rivers

Upnor sands from the west
overlap the Thanet sand beds

Hot desert land surface

Upnor beds cemented by silica

Mangrove swamps

London Clay Sea sinking of the
North Sea Basin

– lithostratigraphy. Geological exploration beneath the North Sea and into north-west Europe has aided correlation. In the centre of the North Sea Basin the rock layers were deposited continuously and so the sequence of rocks there is more complete.

Beneath London, the London Clay layers are riddled with tunnels and shafts of the Tube railway and many other services. Because of this, the more recent tunnelling projects have to be planned to drive through deeper rock layers. The exploration and geological understanding of these deeper layers have been crucial to planning and construction. This has also greatly increased our knowledge of what lies under Essex.

◄ The 5-million-year sequence of events as the land tilted from the north-west (top) with the rifting of the North Atlantic. The Chalk hills continued to rise and the Thanet Sand sea was succeeded by the Upnor Sand sea, with layers of flints eroded from the chalk. Hot desert conditions led to sands and flints being turned into hard masses of 'silcrete' – sarsen and puddingstone. Eventually, the sea returned and deposited thick, clay-rich layers (bottom).

▼ Cores from Thames Tideway Tunnel site investigation show the variable nature of the rocks of the Lambeth Group. *Tim Newman*

The Thanet Sea

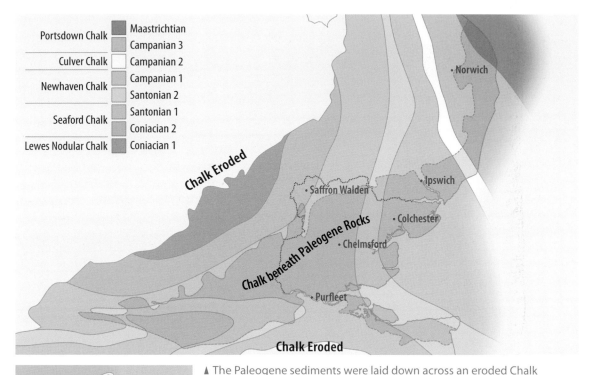

Portsdown Chalk	Maastrichtian
Culver Chalk	Campanian 3
	Campanian 2
Newhaven Chalk	Campanian 1
	Santonian 2
Seaford Chalk	Santonian 1
	Coniacian 2
Lewes Nodular Chalk	Coniacian 1

▲ The Paleogene sediments were laid down across an eroded Chalk landscape. Because the lands had been tipped upwards, erosion removed more Chalk in the north-west. The map shows the successive layers of Chalk beneath Essex, with thicker, less eroded Chalk layers towards the North Sea Basin. *From a map by Andy Gale*

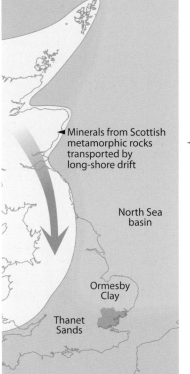

◄ Britain 59 million years ago, showing the drift of sediment from as far as Scotland to form the Thanet Sand layer across Essex, with finer clay sediments reaching the North Sea Basin into East Anglia.

As the opening of the Atlantic continued, the uplifted Chalk and older rocks beneath it were rapidly eroded. The Chalk surface was eroded and dissolved by rain and groundwater.

Great quantities of fine sand were brought by eastward-flowing rivers into shallow seas covering eastern England including Essex. They form up to 30 metres (100 feet) of Thanet Sand.

The Thanet Sand layer forms the lowest part of the 'Lower London Tertiaries' and it rests directly upon the Chalk. In many places it fills solution hollows in the chalk surface. Finer grained silt and mud sediments were carried further out and deposited within the North Sea Basin, where the beds are called the Ormesby Clay. There was also sediment input from the uplands of Scotland which is shown in the analysis of the Thanet Sand taken from boreholes. Minerals typical of rocks of metamorphic origin, such as epidote, garnet and hornblende, were carried south along the coast by longshore drift and incorporated into the Thanet Sand.

The Thanet Sand contains the mineral glauconite, which indicates that it was deposited in a marine environment. This iron-rich mineral is green where it remains buried but, upon exposure at the surface, its iron content weathers to a rusty colour making the sand yellowish-brown. The junction between the Chalk and the Thanet Sand is marked by a distinctive layer of green-coated flints in a sandy clay – the 'Bullhead Bed'. The Bullhead Bed is so called because the flints it contains are not rounded but still have protrusions like horns, just as they came out of the Chalk.

Interesting solution hollows seen in the top of the Chalk at Chafford Gorges are lined with Bullhead flints and infilled by Thanet Sand. Their development has not yet been satisfactorily explained, but they were probably formed after the Thanet Sand was deposited. The flints form a continuous, even layer rather than a pile at the bottom of the pit. The development of these hollows was most likely due to water seeping through the overlying Thanet Sand and down into the Chalk. The water very slowly dissolved out the cracks in the chalk and the overlying flints of the Bullhead Bed plus the sand were let down gradually into the developing hollow. Larger-scale sand-filled hollows can cause difficulties in infrastructure projects that penetrate the boundary between these sands and the Chalk, especially in south Essex.

▼ A solution hollow in the top of the Chalk at Chafford Gorges, lined with Bullhead flints and infilled by Thanet Sand, with a close-up of the 'horned' Bullhead Bed flint lining.

▲ A green-coated Bullhead Bed flint in the wall of North Stifford Church.

A 'bull's head' flint from the Bullhead Bed at Chafford Gorges. ►

The Thanet Formation is thickest in south Essex where it is seen overlying the Chalk in old chalk quarries near Grays and Purfleet. It has been found at depth in many boreholes further east and north where it is covered by younger sediments. These boreholes show that the sands become thinner and more silty northwards and grade into the Ormesby Clay towards the centre of the basin of deposition, where the North Sea is now. Thanet Sand is also seen at the northern edge of the London Basin in the old chalk pits near Sudbury, but they thin westwards and are overlapped by beds of the later Upnor Formation. The Thanet Sand tends not to have any small-scale layering because the sands were mixed by animal activity (bioturbated) during and after deposition. Fossil molluscs, bivalves and rare gastropods indicate a cool water environment less than 50 m (160 ft) deep.

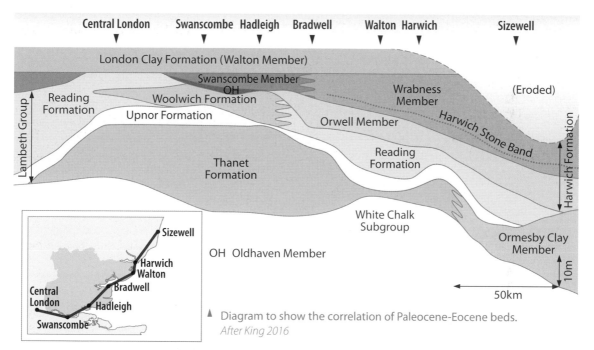

▲ Diagram to show the correlation of Paleocene-Eocene beds.
After King 2016

► Thanet Sand above the Chalk in Sandy Lane Quarry, along the northern edge of the London Basin near Sudbury.
British Geological Survey
© UKRI 2021

The Lambeth Group

Rocks of this Group lie above the Thanet Formation. South-eastern Britain, including Essex, was still on the edge of a shallow sea centred on the North Sea Basin. However, as the North Atlantic was opening out to the west, magma was injected within the crust beneath Britain. The great Cleveland dyke swarm was formed across northern England and the shape of the basin was altered as the crust buckled. The source of sand from the north, including the metamorphic minerals from the Scottish uplands, was largely cut off. The top layers of the Thanet Sand were eroded and weathered prior to the deposition of the following layers of sands and clays, which are therefore placed in a separate geological Group – the Lambeth Group.

The name of the Lambeth Group unit replaces that of the former Woolwich and Reading Beds, including the Woolwich/Reading Bottom Bed. Despite only being a maximum of 15–20 m thick in the London area, these sediments are linked with many geological engineering risks. As the London Clay layer beneath London and into Essex is congested by so many tunnels, more recent projects have been forced into these lower, more complex, varied and problematic rock strata. Detailed studies associated with major underground engineering projects, such as Crossrail and the Thames Tideway Tunnel, have clarified the nature and sequence of these interfingering sand, silt, clay and pebble deposits. This has, in turn, influenced operations during these complex engineering works, demonstrating the crucial importance of detailed geological knowledge.

As there are very few permanent exposures of the rocks of the Lambeth Group, our knowledge of the nature and distribution of these rocks has been

▼ The varied beds making up the Lambeth Group and the Harwich Formation lie above the Chalk and Thanet Formation and beneath the London Clay. *After King 2016*

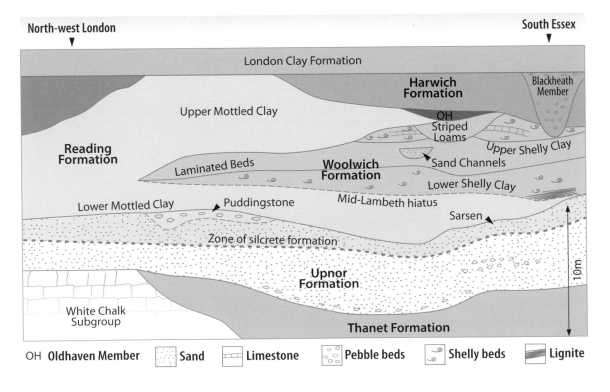

North-west London   South Essex

London Clay Formation

Harwich Formation — Blackheath Member

Upper Mottled Clay

OH

Striped Loams

Upper Shelly Clay

Reading Formation

Laminated Beds

Woolwich Formation

Sand Channels

Lower Shelly Clay

Lower Mottled Clay — Puddingstone — Mid-Lambeth hiatus

Sarsen

Zone of silcrete formation

Upnor Formation

10m

White Chalk Subgroup

Thanet Formation

OH **Oldhaven Member** · Sand · **Limestone** · Pebble beds · **Shelly beds** · Lignite

▲ Buckingham Hill Sand Quarry near Orsett, showing magnificent exposures of the sands of the Woolwich Beds. *P.R. Harvey*

obtained largely from temporary sections such as the M11 and M25 motorway cuttings where a team of amateur geologists carried out much work. Amateur geologists have an important role in discovering and recording such sections before they are lost forever. Without their detective work we would know a lot less about the landscape and wildlife of Essex during this early part of the Paleogene Period.

The Lambeth Group comprises the Upnor Formation (the former 'Bottom Bed'), the overlying Woolwich Formation and the laterally equivalent Reading Formation. The complexity of the relationship between these beds is shown in the diagram.

The sea returns, Upnor Formation

The top of the Thanet Sand was exposed on land and was burrowed and channelled prior to the next advance of the sea. The Upnor Formation, deposited as sea level rose again, extends further west than the Thanet Formation deposits. It overlaps onto the Chalk, indicating that the sea was greater in extent or that a considerable amount of Thanet Sand was eroded away.

Several large sand quarries were worked in the Thanet Sand and the Upnor Formation in south Essex. Most notable are Orsett Cock Quarry (also known as Southfields Quarry), Buckingham Hill Sand Quarry also known as Orsett Tarmac Pit, and the Arena Essex Sand Pit which contains some very large sarsens. Millwood Sand Cliff and Sand Martin Cliff, now part of the Chafford Gorges Nature Park, still retain excellent sections through the Thanet Sand.

In the south of Essex deposits of the Upnor Formation (formerly called the Woolwich Bottom Bed) consist of a pebble bed and green silts and sands with pebble seams, which pass up into yellow sands laid down in a lagoon. These sands are coarser than the Thanet Sand, with a mineral content dominated by zircon with rutile and tourmaline from granite rocks to the south-west. This distinguishes them from the Thanet Sand and indicates their different origins. Deep channels at the margin of the Upnor sea cut into the top of the Thanet Sand. These occur in the Orsett area, where they are filled with alternating layers of fine sand and lenses of rounded flint pebbles, the Upnor Pebble Gravel. Most of the flint pebbles are black on the outside and internally varied brown, grey and red. The black coating on the flint pebbles is characteristic of pebbles derived from the Chalk and then rolled on the seabed and beach for a considerable time. It is thought that the coating is due to the

Britain 57 million years ago, showing the altered coastline and areas of sand and clay. The sea extended further across to the west and Essex was covered by new sand layers – the Upnor Formation.

Dyke swarms indicating injection of magma causing uplift and cutting off northerly sediment supply

Upnor Formation

presence of manganese and originated in the intense global heating events of early Paleogene times. Small crescentic 'chatter marks' on pebble surfaces indicate that they have collided in a high-energy environment such as a storm beach.

▲ Detail of the Woolwich Beds in Orsett Depot Quarry. Cutting through the bedding vertically is the sand-filled burrow of a shrimp.
British Geological Survey © UKRI 2021

◄ Upnor Pebble gravel fills a channel in the top of the Thanet Sand, Orsett Depot Quarry.
British Geological Survey © UKRI 2021

Engineers tunnelling through the Upnor Formation faced a very serious hazard as the air was unexpectedly depleted in oxygen. The presence of green glauconite in unexposed sediments was blamed at first, but after intensive research it was concluded that it was caused by the presence of 'green rust' that coats individual sand grains. This very thin layer of complex iron hydroxides is seen as a bluish-green layer in rock cores, but the colour disappears in a matter of minutes on exposure, as it rapidly strips oxygen from the air. Where this sediment is likely to be encountered during tunnelling works, very great care needs to be taken during construction to avoid the creation of anoxic conditions.

▲ Green, glauconitic Upnor Formation deposits in a Thames Tideway shaft.
Tim Newman

◄ Pebbles in the top of the Upnor Formation, excavated in the Thames Tideway Tunnel.
Tim Newman

Climate, sarsens and puddingstone

There is evidence of a time gap at the top of the Upnor Formation, before the overlying Reading Formation was deposited. During this time, the uppermost sand layers of the Upnor Formation were exposed as a land surface. Iron staining and fracturing, with carbonate and silica cement within the uppermost layers of the sands indicate exposure on a land surface during arid conditions. Soil-forming processes – pedogenesis – penetrating to a depth of 1–2 metres, have been observed in borehole cores and it is these that were probably responsible for the formation of sarsens and puddingstones, collectively termed silcretes.

The climate of Essex was extremely hot and dry, associated with high heat events – notably the Paleocene-Eocene Thermal Maximum (PETM) about 56 million years ago – that have been recorded in ocean bed core sediments. During that time, water with dissolved silica was drawn to the surface and the sands and pebble beds became cemented by silica, in the form of quartz, to form tough sandstones and conglomerates. In modern environments where these processes can be observed happening today, such as in the Australian outback, the mobilisation of silica is in underground channels where the right chemical conditions occur.

◄ Sarsen grading into puddingstone, Stanway near Colchester.

▼ Global climate heating events in Paleocene and Early Eocene times, detected from evidence in ocean sediments.

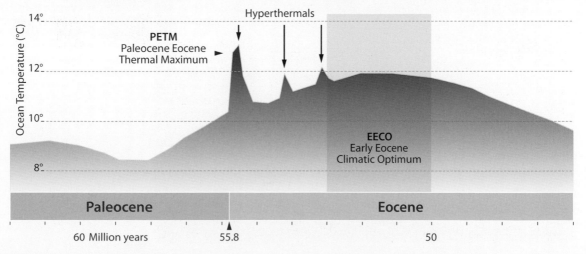

Sarsen stone

▲ Both the top and the base of this sarsen slab show mammillated surfaces. Chafford Gorges near Grays.

The silica-cemented sandstone boulders are known as sarsens, a name that originated in Wiltshire where they occur on the chalk downland (many of the massive stones of Stonehenge are sarsens). They are very common in Essex and are often seen set in the ground as landmarks or built into walls. Puddingstone can be seen to grade into sarsen in some blocks just as pebbles and sand on a modern beach are interbedded.

Large sarsens often have rootlet holes and mammillated surfaces. Rootlet holes indicate further their formation within a soil layer. Mammillated surfaces are particularly well preserved in the sarsens found on the eroded chalk surface in south Essex near Grays, which are very locally derived, so the surfaces are fresh. These rounded concentric features show how the original silica cementation spread from centres of crystallisation within the sandy layer.

Subsequently, when the overlying layers were subjected to erosion, these large blocks of hard rocks were left on the surface. Much later, during the ice age, many blocks moved down slope by freeze–thaw processes – solifluction – and were then picked up and transported by the large ice-melt rivers such as the Thames system and, in parts of Essex, by the ice sheet during the most extensive glaciation, the Anglian, 450,000 years ago.

▼ Sarsen with rootlet holes, Alphamstone churchyard.

Puddingstone

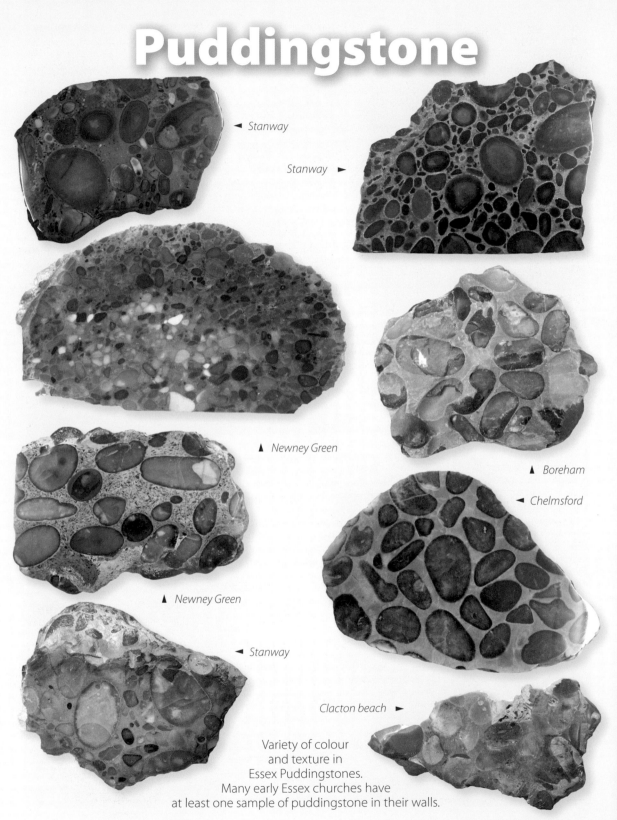

◀ *Stanway*

Stanway ▶

▲ *Newney Green*

▲ *Boreham*

◀ *Chelmsford*

▲ *Newney Green*

◀ *Stanway*

Clacton beach ▶

Variety of colour
and texture in
Essex Puddingstones.
Many early Essex churches have
at least one sample of puddingstone in their walls.

▲ Puddingstone with sarsen in the foundations of North Stifford church.

Puddingstone – its name suggests its resemblance to a plum pudding – consists of pebbles and sand cemented by silica. Silica-cemented sand without pebbles is called sarsen. Few in situ exposures of puddingstone have been found, mostly in Hertfordshire, but these fit the picture of being silica-cemented Upnor Pebble Gravel that was a beach deposit at the margins of the Upnor Formation sea. Observation of the many reworked blocks of puddingstone found in the Quaternary Thames Gravels of Essex show a variety of colours and textures, which suggests that cementation was in discrete blocks and lenses rather than one continuous horizon. It is made up of well-rounded flint pebbles up to 5 cm (2 inches) in diameter in an extremely hard quartz-cemented matrix. The pebbles are often black coated and frequently show crescentic chatter marks.

There are a few puddingstones in south Essex, notably in the walls of North Stifford church. These are locally derived from the Upnor Pebble Gravel deposit and the matrix is somewhat less tough.

Puddingstone should not be confused with other cemented gravels; in the case of puddingstone the pebbles and the enclosing matrix are the same hardness, therefore when a piece of puddingstone is broken, the plane of fracture usually passes through rather than around the pebbles. Some specimens show fine pebbles in the matrix distributed around the larger pebbles. Others have angular shards of broken flint that seem to flow around the pebbles. Flints might have been broken in this way in the hot arid climate and then incorporated into the pebble deposit, but this texture has yet to be fully explained. In some cases, the pebbles are suspended in a very fine matrix where the sand has been partly or fully recrystallised into an extremely tough cement.

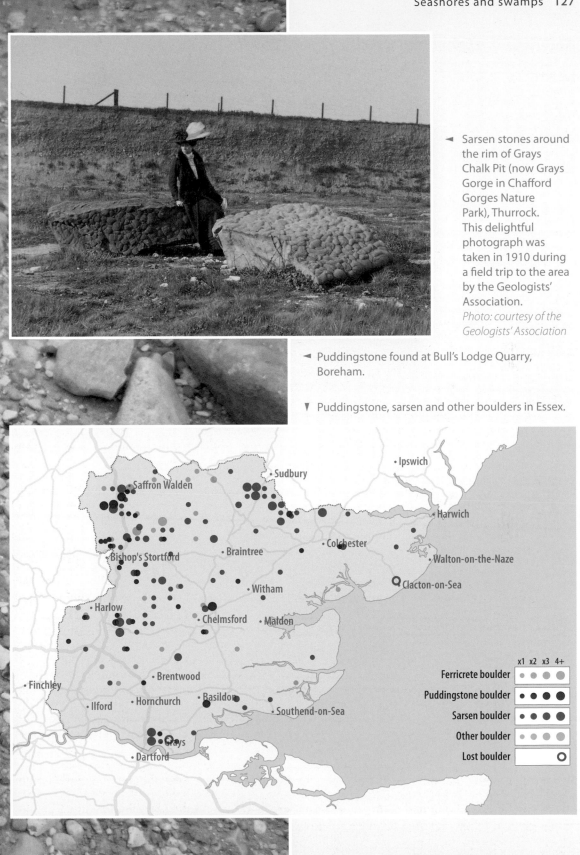

◀ Sarsen stones around the rim of Grays Chalk Pit (now Grays Gorge in Chafford Gorges Nature Park), Thurrock. This delightful photograph was taken in 1910 during a field trip to the area by the Geologists' Association.
Photo: courtesy of the Geologists' Association

◀ Puddingstone found at Bull's Lodge Quarry, Boreham.

▼ Puddingstone, sarsen and other boulders in Essex.

• Ipswich

• Sudbury

Saffron Walden

• Harwich

• Bishop's Stortford

• Braintree

• Colchester

• Walton-on-the-Naze

Clacton-on-Sea

• Witham

• Harlow

• Chelmsford

• Maldon

• Finchley

• Brentwood

• Ilford

• Hornchurch

• Basildon

• Southend-on-Sea

• Grays

• Dartford

	x1	x2	x3	4+
Ferricrete boulder	•	•	•	•
Puddingstone boulder	•	•	•	•
Sarsen boulder	•	•	•	•
Other boulder	•	•	•	•
Lost boulder				○

Return of the Edgelands: Woolwich and Reading Formations

After a period of exposure at the surface when soil-forming processes affected the upper layers of the Upnor Formation, a series of sands and clays was laid down at the margin of a sea which periodically spread westwards across Essex from the North Sea Basin. Beds of the Reading Formation were deposited in the non-marine environment of rivers and lakes and the Woolwich Formation beds of shelly sands and clays were laid down when the sea encroached.

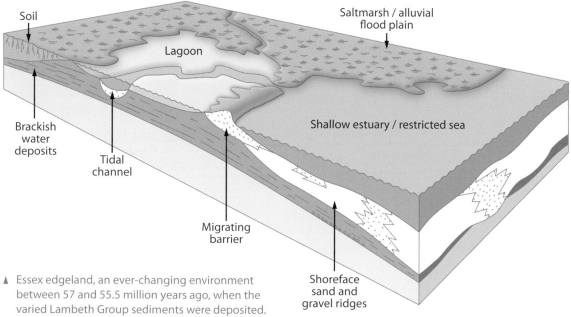

▲ Essex edgeland, an ever-changing environment between 57 and 55.5 million years ago, when the varied Lambeth Group sediments were deposited. *After RA Ellison, BGS*

▲ Gastropod *Brotia melanoides* from the Woolwich Beds at Aveley. *Gerald Lucy*

Subsidence continued through this time which enabled the land-based sediments in particular to be preserved. The sea came and went across the area of Essex. Overall, the Woolwich and Reading Formations are 10–15 metres thick in Essex. Some surface outcrops occur in the south of the county and in the north and north-west between the outcrops of the Chalk and the London Clay. Most of the mapping of the unit requires the digging of trial pits and much more information has been gained from boreholes where these friable sediments are less affected by alteration and weathering.

Reading Formation

The Lower Mottled Clay of the Reading Formation rests directly on the Upnor Formation across most of Essex. The boundary is sometimes difficult to place precisely even in borehole cores. This is because soil-forming processes in the warm, subtropical environment continued when the sediments were above water and caused clay particles to be washed downwards into the top of the Upnor Formation sediments along with the colouring produced by iron minerals.

Much of the Reading Formation sediments consist of mottled or multicoloured silty clay and clay, which was deposited in lakes and river valleys. This characteristic lithology – the type of sediment – was formerly called the 'Reading Beds' or 'plastic clay'. Colours include pale brown, pale grey-blue, dark brown, pale green, red-brown and crimson, depending on the oxidation and hydration state of the iron in the sediments. The red is the oxidised iron mineral haematite, which is commonly formed during drier periods; the yellow is goethite, a hydrated iron oxide that forms in damper conditions.

The Reading Formation clays contain numerous fissures, which give it a blocky texture. The top part contains irregularly shaped nodules, varying from soft and powdery to hard, limy 'calcrete' – ancient soil – up to 0.5 m in diameter. Sands become increasingly dominant to the east and are sometimes referred to as 'Lower Mottled Sand'. Animals, such as burrowing crustaceans and worms, and plants roots all radically changed the texture of the deposits when they were part of the soil on the floodplain, forming the blue-grey to brown coloured network seen in core samples. The activity of bacteria during the decaying of the roots produced grey-coloured rather

▲ Reading Formation, Lower Mottled Clay in borehole cores from the St Pancras area of north London.
Jackie Skipper

◄ Upper Mottled Clay in the Reading Formation, Thames Tideway Tunnel.
Tim Newman

than orange iron minerals. The pause in sedimentation when soil processes were dominant before the sea flooded from the east is called the 'mid-Lambeth Hiatus'. This is a particularly significant layer for engineering geologists planning tunnels and other major construction projects as it is often marked by the presence of very strongly cemented beds – either siliceous or calcareous – that formed as 'hard-pans' within the soil. These can often cause construction difficulties.

Woolwich Formation

The Woolwich Formation rests on the Lower Mottled Clay/Sand of the Reading Formation. It thickens south-eastwards in Essex and is absent in north Essex. It was deposited as the sea flooded the area from the east, overlapping onto the Reading Formation deposits. The marine Woolwich Formation deposits thus form a layer, thinning to the west, within the non-marine Reading Formation. These beds contain fossil pollen grains showing that deposition was in the brackish waters of tidal channels in a river estuary.

Characteristic of fluctuating Essex edgeland environments, the temporary rise in sea level led to the establishment of lagoon and estuarine conditions and the deposition of the Lower Shelly Clay and the Laminated Beds in the southern and eastern parts of the county. Sand beds formed in channels that straddled the flood plain, cutting into the sediments below. A project at the scale and depth of Crossrail can cut into some areas where sand channels are more common. These are a hazard to tunnelling as they often contain over-pressured water. They frequently occur along fault alignments where the land surface was more easily eroded and where rivers tended to naturally pick out these weaknesses.

▲ The thinly bedded and burrowed beds revealed in the side of a swallow hole at Hill Farm, Gestingthorpe in north Essex are probably Reading Formation.
Ashley Cooper

Close-up view of the ▼ lignite band.
Essex Field Club

The lower boundary of the Lower Shelly Clay is sharp and well-defined on the often pale and bleached Lower Mottled Clay which had been affected by soil formation. Burrows containing dark clay, lignite or shells extend 1.5 m (5 ft) into the Lower Mottled Clay. The main rock type is dark grey to black clay that contains abundant shells. These shell-dominated beds suggest that sediment input was slow, thus allowing the development of shell banks. Bivalves such as the oyster *Ostrea bellovacina* and gastropods such as the whelk *Brotia melanoides* occur, indicating estuarine conditions. Freshwater gastropods (snails) in thin seams in the shelly clays have been found at Aveley and Leytonstone; these indicate temporary freshwater conditions on the dominantly brackish mudflats. Lignite is commonly seen at the base of the Lower Shelly Clay, especially in south Essex where a considerable thickness of this black peaty material was revealed in cuttings for the M25 at Aveley and in the HS1 cutting in north Kent. This lignite is the remains of flooded vegetation that had been growing around the lakes and rivers.

Northwards and eastwards in Essex, the Woolwich Beds are overlain by the more terrestrial Upper Mottled Clay of the Reading Formation. Eventually the Woolwich Beds pinch out altogether, indicating the extent of the sea incursion. Deposition of the Lambeth Group sediments was followed by a period of uplift and retreat of the shoreline to the east with later erosion.

▼ Black band of lignite at the base of the Lower Shelly Clay in the Woolwich Beds at Aveley exposed in works for the M25 in 1978. *Essex Field Club*

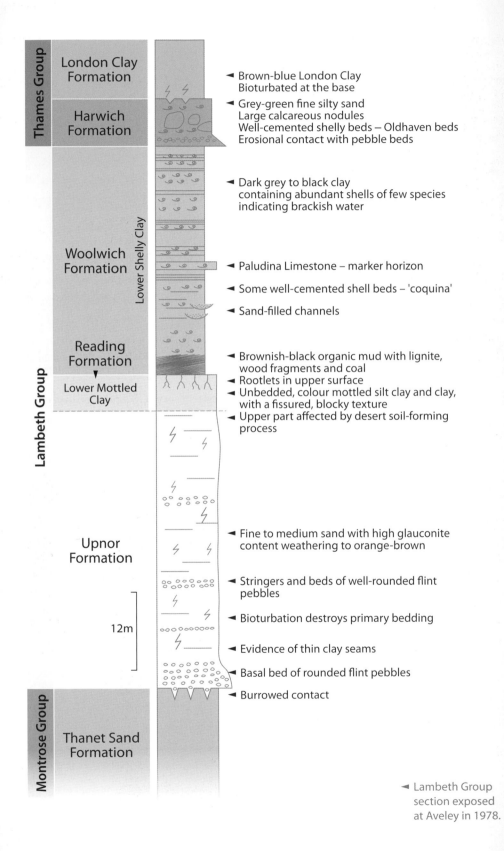

Thames Group

London Clay Formation
- ◄ Brown-blue London Clay Bioturbated at the base

Harwich Formation
- ◄ Grey-green fine silty sand Large calcareous nodules Well-cemented shelly beds – Oldhaven beds Erosional contact with pebble beds

Lambeth Group

Woolwich Formation

Lower Shelly Clay
- ◄ Dark grey to black clay containing abundant shells of few species indicating brackish water
- ◄ Paludina Limestone – marker horizon
- ◄ Some well-cemented shell beds – 'coquina'
- ◄ Sand-filled channels

Reading Formation
- ◄ Brownish-black organic mud with lignite, wood fragments and coal

Lower Mottled Clay
- ◄ Rootlets in upper surface
- ◄ Unbedded, colour mottled silt clay and clay, with a fissured, blocky texture
- ◄ Upper part affected by desert soil-forming process

Upnor Formation
- ◄ Fine to medium sand with high glauconite content weathering to orange-brown
- ◄ Stringers and beds of well-rounded flint pebbles
- ◄ Bioturbation destroys primary bedding

12m
- ◄ Evidence of thin clay seams
- ◄ Basal bed of rounded flint pebbles
- ◄ Burrowed contact

Montrose Group

Thanet Sand Formation

◄ Lambeth Group section exposed at Aveley in 1978.

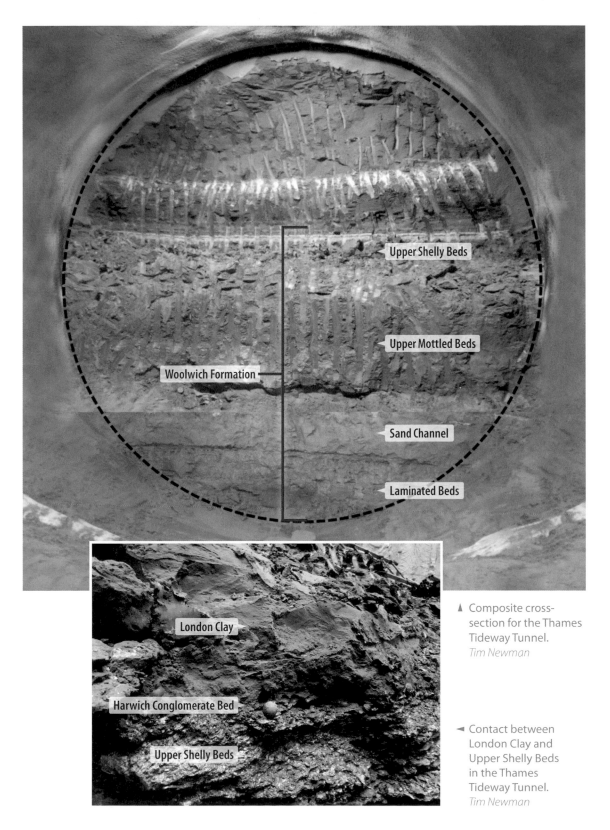

Upper Shelly Beds

Upper Mottled Beds

Woolwich Formation

Sand Channel

Laminated Beds

London Clay

Harwich Conglomerate Bed

Upper Shelly Beds

▲ Composite cross-
section for the Thames
Tideway Tunnel.
Tim Newman

◄ Contact between
London Clay and
Upper Shelly Beds
in the Thames
Tideway Tunnel.
Tim Newman

Pebble beds and shell banks
Thames Group: Harwich Formation

Once more the sea flooded across the land from the east, with a shoreline to the south and west of Essex. Near to the shore, more sands, gravels, silty sands and silty clays were deposited in thin, discontinuous layers. In some places eastward-flowing rivers formed deep, gravel-filled channels. Elsewhere, shell banks were heaped up and eventually cemented to form hard shelly layers. These are called 'coquinas' when encountered in engineering works, as they form hard bands that require drilling to break them up. Towards the centre of the basin, layers of mud accumulated, with periodic additions of volcanic ash blown from the north-west from volcanoes associated with the opening of the Atlantic. These muddy sediments form a continuous sequence with the overlying London Clay and, as they are indistinguishable from the London Clay in the field, they will be described with these deposits in the next chapter.

Blackheath Member

The Blackheath Member is a series of pebble beds which occupy channels that cut right through the underlying Woolwich and Reading Formations. They occur in south-east London and Kent and may be present in the extreme south-west corner of Essex, but are not exposed at the surface. The well-rounded, black-coated pebbles, reworked from earlier marine pebble beds, give Blackheath in south-east London its name. The most famous Blackheath Beds locality visible today is the sand pit at Abbey Wood in south-east London which has yielded a large number of fossils. As well as where the Blackheath Beds occur at the surface, most patches of sands and gravels laid down in the early Paleogene Period are associated with heathland and acid soils.

Oldhaven Member

After another break in sedimentation, accompanied by erosion in places, the beds of Oldhaven Member were deposited in channels cut down into beds of the Lambeth Group or the Blackheath Formation. They are mostly sands, but also contain well-rounded flint pebbles, especially at the base. They crop out in the south of Essex, but are exposed only in temporary excavations where they are found to be rarely more than 3 m (10 ft) thick. In places they contain shell seams which yield a distinctive marine fossil assemblage. Bivalves such as the cockle *Glycymeris plumstediensis*, gastropods such as the whelk *Pseudoliva fissurata*, and the teeth of the sting ray *Hypolophus sylvestris* are typical Oldhaven Bed fossils. The Oldhaven Beds also contain an abundance of shark and fish teeth and bones. Parts of the Oldhaven Beds are cemented by calcium carbonate into hard masses of sandstone which make the more fragile shells easier to recover intact, but any unexpected hard bands such as these may cause problems in engineering projects.

A temporary section through the Oldhaven Member was exposed in excavations for the M25 motorway at Aveley where, although less than 1 metre (3 ft) thick, the richly fossiliferous shell seams yielded over 60 species, including a bone of the earliest bird to be found in Essex. In the north of the county the lateral equivalent of the Oldhaven Member (formerly called the London Clay Basement Bed) consists of fossiliferous silty clays. These clays crop out in the north-west and north of Essex and were seen in 1979 in a temporary pit adjacent to the M11 motorway at Elsenham.

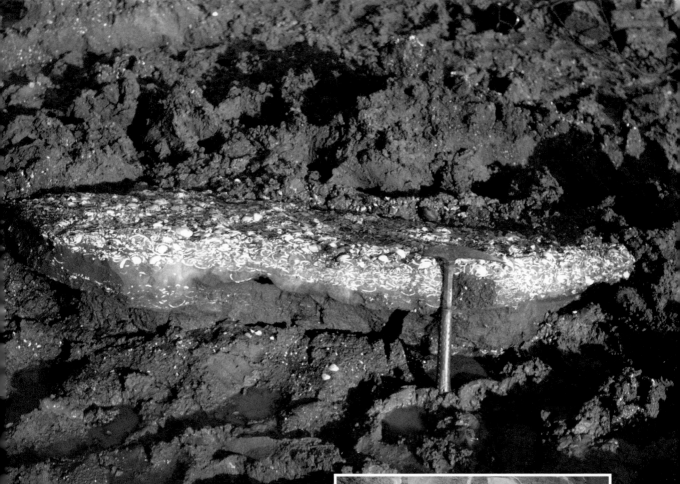

▲ Fossil shells in a hard band of Oldhaven Beds excavated during the building of the M25 motorway. *Essex Field Club*

Detail of a block of shelly sandstone from the ▶ Oldhaven Beds at Aveley. *Gerald Lucy*

A pebble bed, known as the Suffolk Pebble Bed, represents a beach as the sea covered more and more of the land. This pebble bed underlies north-east Essex and is exposed along the River Deben in Suffolk where it yields sharks' teeth, molluscs and rare mammal remains. Crustal tilting and subsidence in the North Sea Basin continued and Essex was slowly being more deeply submerged beneath the London Clay sea – the subject of the next chapter.

7

Grey layers of London Clay are exposed on the foreshore and in the cliffs at Walton-on-the-Naze. Shark teeth weather out of the clay and can be found along the shoreline. Fossil wood and other plant remains are found here, too.

	Paleogene		Neogene	
66 million years	55.5 51.5 49		23	2.58 Now

▲ London ▲ ▲
Clay   Delta

Palm trees and crocodiles

Subtropical Essex

About 55 million years ago, the area of Essex was eventually submerged beneath a warm sea up to 200 m (650 ft) deep. Muddy sediments were deposited across south-eastern England, with palm trees on land and crocodiles in the sea showing that there was a subtropical climate. During this period, Essex was situated around latitude 40° north, the same as present-day Naples.

The North Sea area was continuing to subside and tilt but, by this time, the volcanic activity to the north-west – associated with the opening of the Atlantic Ocean – was beginning to wane and land erosion decreased as episodes of uplift declined. The sea increased in extent and the area of Essex was completely submerged and became less affected by changes in the position of the coastline to the west. Consequently, sedimentation was more uniform for longer periods of time. The coastal area, which could have been as far away as the Midlands, supported a rainforest and mangroves. Rivers flowed into this sea bringing mud and silt which settled and became compacted. Eventually, a thickness of up to 200 m (650 ft) of blue-grey clay, generally known to us as 'London Clay', was built up on the seabed.

Wherever it is seen, the whole of the London Clay appears at first glance to be uniform. Detailed studies have in fact shown how the layers vary and help us to understand how they were deposited. Boreholes were drilled by the British Geological Survey (BGS) in the 1970s and subsequent studies clarified the pattern of sedimentary cycles caused by sea level changes. This study revealed that the lowest clay beds were deposited at the same time as the Blackheath pebble beds and the shelly Oldhaven beds described in the previous chapter. Because of this time-equivalence, the whole of the lowest part of the sequence – the clay, limestone and gravels – is now classified by BGS as the Harwich Formation, and not part of the London Clay Formation. The base of the Harwich Formation is younger to the west because the sea deepened over time and the shoreline gradually extended westwards. Because the clays of these two Formations are indistinguishable without detailed analysis, they are both commonly referred to simply as London Clay.

◄ Swamp coastline in eastern USA today, similar to Britain in London Clay times.
Photo by Matt Rath/ Chesapeake Bay Program

▲ Geography of Britain from 55 million years ago when the London Clay was deposited. The climate was subtropical and the coastline thickly forested. Rivers brought mud and silt into the shallow sea that covered south-east England. The coastline gradually moved south-eastwards as the deposits accumulated over a 3-million-year period.

Eocene rocks beneath Essex

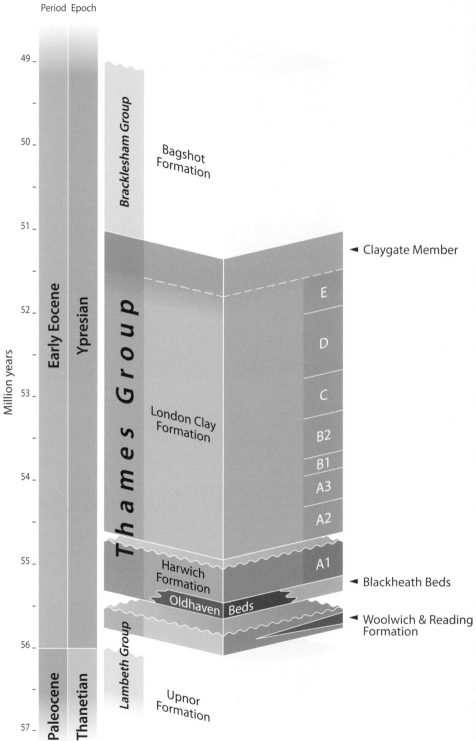

Period Epoch

Million years

49 —

50 —

51 —

52 —

53 —

54 —

55 —

56 —

57 —

Early Eocene — Ypresian

Paleocene — Thanetian

Bracklesham Group

Thames Group

Lambeth Group

Bagshot Formation

London Clay Formation

Harwich Formation

Oldhaven Beds

Upnor Formation

◄ Claygate Member

E

D

C

B2

B1

A3

A2

A1

◄ Blackheath Beds

◄ Woolwich & Reading Formation

Harwich Formation
Harwich Stone Band

Several hard beds of cementstone occur near the base of the cliffs at Wrabness, Harwich and farther around the coast to the south. These formed when the clay became cemented by lime-rich mud, probably from accumulations of shells when deposition was less active. Cementstone often has cracks filled with large white calcite crystals.

Bands of cementstone are well exposed along the River Stour on the foreshore and on the beach at Harwich. The most prominent bed is called the Harwich Stone Band. This was exploited to make cement from Roman times when it was collected and broken into small pieces and heated in kilns. The industry persisted into the nineteenth century before it was superseded by Portland cement. Such was the demand for cement in the Victorian era that the harbour and cliffs at Harwich came under threat from the quarrying of the cementstone. A government commission prohibited the taking of stone from the cliffs which then led to dredging from offshore. During the process of exploiting this resource, many valuable fossils were found, such as turtles and other marine creatures, and mammals including the ancestral horse *Hyracotherium*.

Close-up of ▲ cementstone in the Harwich Stone Band at Wrabness. Layers containing volcanic ash show up as they weather to a paler colour.

Harwich Stone Band ► on the foreshore at Wrabness. The large cracks were filled with calcite crystals before they were eroded out on the beach.

The cliffs north of Wrabness expose the whole of ► the upper part of the Harwich Formation. *Creative Commons Geograph Britain CC BY-SA 2.0 David Kemp*

▲ Recycled cementstone, probably from a nearby Roman villa, in a Norman church wall, St Mary Broomfield.

◄ Colchester Castle is largely constructed of cementstone. *Wikimedia Commons /A. Walker*

Cementstone was also used extensively as a building stone, particularly by the Romans; Colchester Castle is built almost entirely of cementstone blocks. They were picked up from the shore and carried far inland along the rivers of Essex for the building of many significant buildings such as large villas and temples. These were subsequently plundered by Saxon and Norman church builders and are now seen in many early medieval churches where they weather to a characteristic brownish yellow or gingery colour.

The Harwich Stone Band is about 0.25 m (10 in) thick and includes a central ash layer. It forms conspicuous outcrops around the Essex coast and is a useful marker in boreholes. In offshore seismic reflection surveys, it is a prominent reflector called the Paleocene Ash Marker which has been traced extensively in the North Sea to reveal geological structures during oil and gas exploration.

Volcanic ash bands

Above the Harwich Stone Band, the clay also contains seams rich in volcanic ash. There were active volcanoes associated with the opening of the North Atlantic Ocean between Greenland and north-west Europe at this time and ash was carried eastwards periodically and fell into the sea to be incorporated in the clay beds. The paler, ash-rich bands are particularly well seen in the cliffs at Wrabness and at the northern end of the cliff at the Naze.

▲ A narrow, concentrated ash band at Walton-on-the-Naze.

Close-up of ash ► bands in the cliff at Walton-on-the-Naze.

▼ Volcanic ash bands from around 55 million years ago are seen clearly in the cliffs at Walton-on-the-Naze.

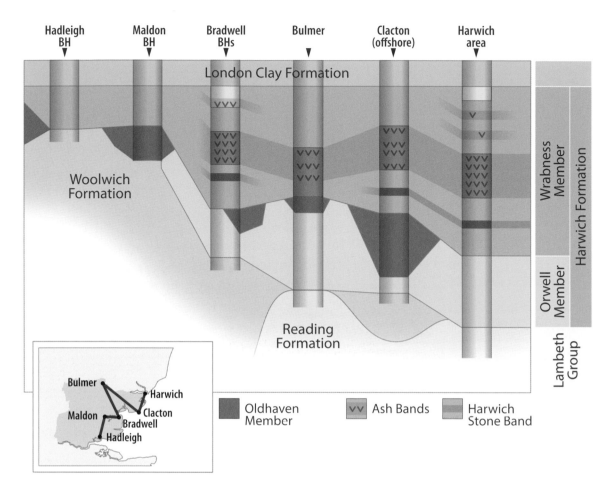

Hadleigh BH   Maldon BH   Bradwell BHs   Bulmer   Clacton (offshore)   Harwich area

London Clay Formation

Woolwich Formation

Reading Formation

Wrabness Member

Harwich Formation

Orwell Member

Lambeth Group

Oldhaven Member   ᵛᵛ Ash Bands   Harwich Stone Band

▲ Correlation of the Harwich Formation lithostratigraphy showing the beds thickening towards the north-east.

Bulmer Brickworks cliff ► showing ash bands.

The London Clay

London Clay underlies much of Essex. It crops out at the surface forming a large clay vale eastward from the Hornchurch area across to Maldon and Southend-on-Sea. Here gardeners will recognise the sticky, malleable clay soil that holds water so well, but which shrinks and cracks and turns very hard in times of drought. The farmland is described as 'three-horse plough country' and chalking was required every five years or so to make the soil workable and more fertile.

◄ London Clay landscape view from Buttsbury towards Ingatestone and Fryerning.

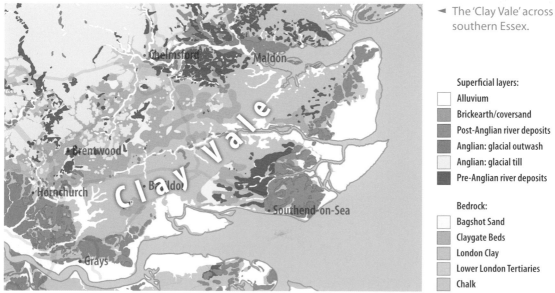

◄ The 'Clay Vale' across southern Essex.

Superficial layers:
☐ Alluvium
■ Brickearth/coversand
■ Post-Anglian river deposits
■ Anglian: glacial outwash
☐ Anglian: glacial till
■ Pre-Anglian river deposits

Bedrock:
☐ Bagshot Sand
■ Claygate Beds
■ London Clay
■ Lower London Tertiaries
■ Chalk

London Clay is best seen along the coast or river-sides. There are few inland exposures except for temporary excavations for engineering works such as major roads. When the M25 was built in the 1980s members of Essex Rock and Mineral Society were able to log and record sections through the London Clay

▲ Cutting for the M25 at Nag's Head Lane, Brentwood.
Essex Field Club

sequence and search for fossils. The M25 cutting at Nag's Head Lane, south of Brentwood, is the largest cutting through London Clay rocks in the London Basin.

The properties of London Clay

London Clay is a stiff blue-grey clay, which weathers to a brownish colour near the surface. It is predominantly composed of a mixture of clay minerals, plus quartz in the more silty layers. Its characteristic 'plastic' nature is due mainly to the clay mineral called smectite. Due to its very large chemically active surface area, up to 700 square metres of surface area per gram of clay mineral, smectite readily absorbs water into its mineral structure. This causes the clay to swell up and become plastic. As it dries out it shrinks again. For this reason, the clay mineral content of London Clay is a critical engineering consideration.

Smectite originates from the weathering of volcanic ash. Within and above the Harwich Stone Band, discrete ash layers were formed when particles from the clouds of ash spewed out periodically by the volcanoes of the rift zone between Greenland and Europe drifted eastwards and landed in the sea. Further west, the ash fell on the land surface and was incorporated into soils. The ash enriched the soils and enabled them to support a rich variety of vegetation. These soils were then inundated by rising sea levels and eroded and carried eastwards into the London Clay sea. In this way, the volcanic smectite was more evenly distributed within the later London Clay beds. The smectite content of the clays decreases towards the top of the beds as the ash-rich soil was eroded away.

The dark grey colour of unweathered London Clay is due to finely dispersed pyrite, iron sulphide. The ultimate source of this pyrite is the anaerobic decay of organic material transported downstream by rivers draining densely forested land to the west. Civil engineers need to take the acid effect of pyrite into account in the specification of concrete for foundations buried in London Clay. Pyritised wood and pyrite nodules weather out of the London Clay and are found in pockets on the beach adjacent to London Clay cliffs. These 'copperas' nodules formed the basis of an early chemical industry dating from the sixteenth century. They were collected and taken to the local Copperas House for processing.

At certain levels the London Clay contains layers of septarian nodules, rounded calcareous cementstone concretions with radiating cracks called septa, often filled with white or yellow calcite. Septaria are very similar in composition to the cementstones in the underlying Harwich Formation. Numerous small concretions of calcium phosphate, phosphatic nodules called 'coprolites', are also common at some levels.

◀ London Clay, a stiff dark blue-grey clay weathering to brown. *Essex Field Club*

▼ Copperas, pyritised fossil wood, on the beach at Walton-on-the-Naze. *Ruth Siddall*

It probably took about 3 million years for all of the London Clay to be laid down. Over this period the sediment was deposited in cycles as the sea level and thus the position of the shoreline fluctuated. Closer to the shoreline, in the Hampshire Basin and west of London, breaks in sedimentation are more distinct than in the deeper water deposits of Essex, which has enabled them to be identified more readily. Each cycle commenced with fine grained clay-rich sediment that was carried out to sea in suspension. The clay minerals then clumped together in the salt water where they settled to the seabed. Coarser sediment from the land gradually built out over the finer clay deposits until sea level rose again.

Only small parts of the London Clay succession can be seen at any one exposure. Cores from boreholes that were drilled by the British Geological Survey in the 1970s at Stock, south of Chelmsford, and Hadleigh, west of Southend-on-Sea, were used to verify the complete sequence.

▲ A present-day mangrove swamp in Cambodia.
Wikimedia Commons/Leon Petrosyan

▼ Fossil wood in London Clay on the foreshore at Walton-on-the-Naze.

The London Clay succession

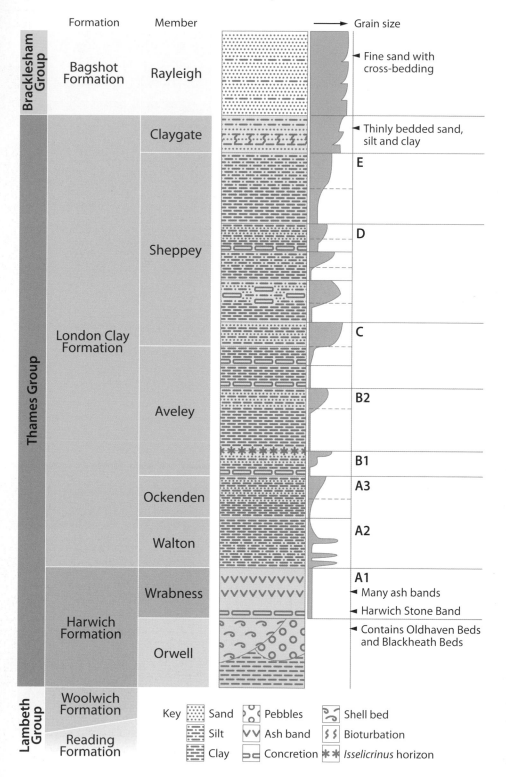

Fine sand with cross-bedding

Thinly bedded sand, silt and clay

Many ash bands

Harwich Stone Band

Contains Oldhaven Beds and Blackheath Beds

Key

Sand	Pebbles	Shell bed
Silt	Ash band	Bioturbation
Clay	Concretion	*Isselicrinus* horizon

London Clay fossils and zones

There are five main cycles, A to E, of subtle 'coarsening up' sequences within the London Clay that can be traced across the whole basin area. Within each sequence, individual fine silt or sand beds a millimetre or so thick were probably the result of distant onshore storms. The resulting floods would have transported masses of vegetation and animal remains into the sea, giving rise to the rich variety of plant and animal fossils found in the London Clay.

The lowermost part of the 'London Clay', division A1, is equivalent to the Harwich Formation. The whole of division A is seen in the cliffs at Wrabness. At Walton-on-the-Naze, the very top of division A1 and division A2, the bottom of the London Clay Formation, are seen in the cliff and foreshore exposures from which shark teeth and pyritised fossil wood are eroded.

The beach at the Naze is currently the best site in Essex for collecting shark teeth and other London Clay fossils. The cliffs with London Clay at the base are being rapidly eroded and the shark teeth in particular are concentrated by wave action in gravel pockets on the beach.

▼ Searching for shark teeth on the London Clay foreshore at the Naze.

◄ Your prize!
A shark tooth in the Naze beach shingle.

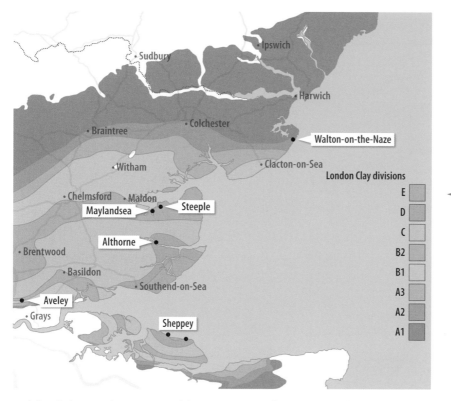

London Clay divisions

E
D
C
B2
B1
A3
A2
A1

◄ The London Clay divisions across Essex, showing classic fossil-collecting localities.

Inland, the very lowest part of the quarry at Aveley went into the upper part of division A3. This quarry exposed almost 50 metres of the lower and middle part of London Clay, which was used for cement manufacture until the 1970s. The pits produced thousands of fossils, including forms new to science, and much research was carried out to work out the interaction between the plants and animals of the time. Measuring the alignment of logs of fossil wood revealed the prevailing current direction on the floor of the London Clay sea.

The whole of division B was exposed at Aveley, but this is no longer visible as the site was landfilled in the 1990s. However, there are various riverside and coastal exposures of both B1 and B2 divisions in Essex, as well as in Kent. The diagnostic fossil for the B1–B2 boundary is a 'sea lily', the crinoid *Isselicrinus subbasaltiformis*. This boundary is

The crinoid *Isselicrinus subbasaltiformis* marks the ► B1–B2 boundary within the London Clay. *Jeff Saward*

exposed at Osea Island where abundant pyritised fragments of *Isselicrinus* occur. Division B2 is well exposed on the foreshore at Maylandsea and Steeple Bay where fine specimens of the lobster *Hoploparia* can be collected from the wave-eroded surface of the clay. Many of the lobster fossils are juvenile moult stages and careful collecting has enabled the seabed of what is now the River Blackwater to be identified as an extensive nursery ground for young lobsters during B2 times.

◄ Aveley Pit, being prepared for landfill operations in 1992. The whole of London Clay Division B and some of Division C was seen here and fossils were collected from it in the 1980s.
David Turner

▲ Careful collecting from the foreshore of the River Blackwater shows that here was a nursery ground for young lobsters *Hoploparia gammaroides*. An adult from Seasalter in Kent for comparison. *Jeff Saward*

▲ London Clay exposed
on the foreshore
at Maylandsea.
Jeff Saward

The surface of a block ►
of cementstone
from the low cliff at
Maylandsea showing
multiple burrows made
by organisms living in
the seabed.
Jeff Saward

Large cementstone nodules that fall from the low cliff at Maylandsea reveal intensely burrowed surfaces, showing the abundance of life forms even though there are no body fossils of the burrowing organisms themselves. The smaller worm burrows are from species known as *Chondrites* and the larger examples are perhaps the burrows of *Callianassa* mud shrimps.

South Ockendon clay pit

The upper part of Unit A and the lower part of Unit B were exposed here and Unit A3 has been named the Ockendon Member and defined from the section here. Fossils such as molluscs, bird bones, fish remains, echinoid spines and turtle bone fragments have all been found. Shark and ray teeth have also occurred, including the tooth of a new species of sting ray, *Dasyatis wochadunensis*, named after Wochaduna, the early English name for (South) Ockendon. The London Clay here has also yielded nine species of plant, found by sieving woody pockets in the clay for fossilised seeds and fruits.

▼ Fossil collecting in South Ockendon Clay Pit in the 1980s. *Essex Field Club*

▼ Drawing of sting ray *Dasyatis brevicaudata*, a species related to *Dasyatis wochadunensis* found at South Ockendon, 1911. *Wikimedia Commons*

Leca Pit, High Ongar

▲ Leca Pit, High Ongar, in 1984. This pit in the London Clay yielded some magnificent fossil specimens, particularly the crab *Zanthopsis* and various species of nautilus. *David Turner*

◄ A fossil crab *Zanthopsis leachii*.
Jeff Saward

Division C is not now well represented in Essex in accessible outcrop, although it was seen at Aveley Pit and at Leca Pit, High Ongar, before they were landfilled. Division D is seen at Butts Cliff between Althorne and Burnham-on-Crouch, where a cliff on an outer bend of the River Crouch is being eroded. Many crabs and thousands of shark teeth from a diverse range of sharks have been collected from here by diligent searchers. Division E is only found in the central part of the London Basin and exposures are mainly in Kent, but there are very limited foreshore exposures in the Southend area. Special techniques and a great deal of care and patience are required when fossil collecting from the London Clay.

Butts Cliff, Althorne

The Cliff or Butts Cliff is a low cliff of London Clay on the outer bend of the north shore of the River Crouch, between Althorne and Burnham-on-Crouch. It is an interesting site, scheduled as an SSSI, which has yielded a remarkable number of London Clay fossils since it was discovered in the early 1970s and it is still accessible along a public footpath today. Finds include fossil fish (mainly shark teeth), crustaceans and bird bones. The fossils have been washed out of the clay by the tide and are found loose in the beach shingle. It is the type locality for several species of fish that have been named for Burnham – *Weltonia burnhamensis*, a small shark, and the ray *Burnhamia daviesi*. Selenite (gypsum) crystals can also be found. Finding fossils at Althorne requires good eyesight and a lot of patience. Searching is usually carried out by lying on the beach, raking through the shingle or sieving shingle for the smaller shark and ray teeth.

▲ *Zanthopsis leachii. Jeff Saward*

▲ *Zanthopsis leachii* in beach single. *Jeff Saward*

▼ Fossil collecting on the foreshore at Butts Cliff.
  Gerald Lucy

▼ Ray tooth *Burnhamia daviesi. UK Fossils*

▼ *Weltonia burnhamensis* cow shark tooth. *David Ward*

Ecology of the London Clay

Fossil collecting from the London Clay in Essex has taken place over the centuries and has led to some remarkable finds. This has enabled a very clear picture of the ecology of the time to be built up. Much dedicated collecting both from foreshore sites and inland quarries and clay pits, particularly when the pits were worked by hand, went on in the past and is continued by enthusiasts today. The whole stratigraphic sequence of the London Clay was examined and fossils were collected during the construction of the M11 from Debden to Chigwell in the late 1970s. At Nag's Head Lane, south of Brentwood, excavation of the deep cutting for the M25 motorway yielded a wealth of specimens to many collectors in 1981, making it the most famous temporary geological site in the county.

The finds show that the London Clay sea was home to turtles, sea snakes, crocodiles and many species of fish including sharks, the largest of which was *Otodus obliquus*, an extinct ancestor of the great white shark. There was also the pearly nautilus, an animal with a spiral chambered shell, the descendants of which now live in the southern Pacific and Indian oceans. Other creatures include gastropods, bivalves, brachiopods, crinoids, crabs and lobsters. High Ongar clay pit yielded some magnificent specimens, particularly the crab *Zanthopsis* and various species of nautilus. South Ockendon clay pit yielded teeth of a new species of stingray and a number of bird bones in the mid-1970s.

An ecological ► reconstruction of the London Clay sea and seabed, based on the Aveley flora and fauna. *After W.S. McKerrow (ed.) 1978*

Fossils from the London Clay

Turricula helix
M25 Cranham

Eopleurotoma prestwichii
M25 Cranham

Turricula teretrium
M25 Cranham

Streptolathyrus zonulatus
M25 Cranham

Argillochelys
Turtle carapace

Fusinus wetherelli
M25 Brentwood

Euthriofusus complanatus
M25 Cranham

Volutospina nodosa
M25 Cranham

Athleta denudatus
M25 Brentwood

Atrina affinis
M25 Cranham

Euspira glaucinoides
M25 Cranham

Fusinus coniferus
M25 Brentwood

Scaphander polysarcus
M25 Cranham

Striarca wrigleyi
M25 Brentwood

Nautilus *Cimomia imperialis*
Roxwell

Deltoidonautilus sowerbyi
M25 Brentwood

Xenophora extensa
M25 Brentwood

Calpitaria sulcatarius
M25 Brentwood

Shark vertebra
Burnham

Sand shark teeth
Walton-on-the-Naze

Otodus obliquus

Striatolamia
macrota

Fish fossils River Blackwater

Crab in nodule
unprepared

Myliobatis
Ray teeth

Rotularia
bognoriensis
Worm tube
M25 Brentwood

Zanthopsis leachii
M25 Cranham

Part of a fish Walton-on-the-Naze

Paracyathus
caryophyllus
Solitary coral
M25 Brentwood

Ditrupa plana
M25 Brentwood

Dromilites belli

Portunites incerta

Isselicrinus subbasaltiformis

Trachysoma scabrum

Isselicrinus subbasaltiformis
River Blackwater

Hoploparia gammaroides (juvenile)
Maylandsea

Hoploparia gammaroides

Important fossils were found, including many fine turtle fossils, when cementstones were collected for the Harwich cement industry in the middle of the nineteenth century. Of particular interest was the first discovery of *Coryphodon*, one of the largest Eocene plant-eating mammals, from a fragment of jaw with teeth dredged up off the coast. From subsequent fossils found in North America, this creature appears to have been almost as large as a rhinoceros, having a large head and knife-like upper canine teeth. Also dredged up was a fragment of the Harwich Stone Band containing the skull and partial skeleton of the earliest known horse, *Hyracotherium*, also known as *Eohippus*. This tiny creature, no larger than a fox, had toes instead of hooves. Its bones, like those of *Coryphodon*, must have drifted downriver on a raft

of vegetation and been deposited on the floor of the London Clay sea. Many more fossil mammals from the London Clay have been found since those days, mostly from Walton-on-the-Naze, where collectors frequently search the beach for specimens. These fossils are of great importance as they provide us with an insight into the evolution of mammals following the extinction of the dinosaurs.

▲ A large oil painting made for the former Geological Museum in South Kensington by E. Marsden Wilson depicting an 'ideal landscape of the London Clay Period'. *Alamy*

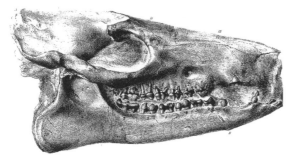

▲ Engraving of the original horse skull, *Hyracotherium* (formerly *Eohippus*), found in the London Clay at Harwich in the early nineteenth century. *Proceedings of the Geological Society of London*

▲ The fossil skull of an ancestral horse – 'Dawn Horse' – collected by Mike Daniels at Walton-on-the-Naze. *Michael Daniels*

▲ A reconstruction of *Hyracotherium* painted by Heinrich Harder (1858–1935). *Wikimedia Commons*

◀ An Eocene mammal. *Wikimedia Commons /Nobu Tamura*

Jawbones from ▶ small insectivorous mammals, Walton-on-the-Naze. *Michael Daniels*

1 cm

▲ Fossil skull of *Coryphodon lobatus*. *Wikimedia Commons/James St. John*

▲ A reconstruction of *Coryphodon* painted by Heinrich Harder (1858–1935). *PD Wikimedia Commons*

The London Clay also contains remarkably well-preserved bones of birds, again mostly from Walton, a site that has produced a large number of finds in recent years owing to the dedicated work of one particular collector, the late Mike Daniels. In fact, the finds from Walton are the best-preserved bird fauna of this age to be found anywhere in the world and therefore the site is of international importance. The fossils are mostly from silty pockets in the lower part of the cliff, probably originally hollows on the ancient seabed, which often contain associated bones of single individuals. The careful separation of the bones from the clay allows the skeletons to be reconstructed. The bones of some 150 possible species have been found including a parrot and a possible ancestor of the ostrich.

Fossil Birds from the London Clay

An amazing number and variety of fossil bird bones have been obtained by dedicated collector, Mike Daniels, from the lower part of the London Clay at Walton-on-the-Naze.

Many of these await formal description and may be new to science.

Fossil bird bones in situ
in the London Clay at Walton-on-the-Naze.

By Eocene times, birds had undergone extraordinary adaptive radiation to produce passerines (perching birds), including all songbirds, and non-passerines, including owls, seabirds, waders, flightless aquatic birds and flightless land birds.

Most of the bird families we see today existed by the end of the Eocene period.

Reconstruction of
Zygodactylus grivensis,
an Eocene passerine.
© Jack Wood

Life reconstruction of
Septencoracias morsensis
a roller-type bird,
Lower Eocene Mo-Clay, Denmark.
Bourdon et al. (2016)

Bones of an *Eopasser*.

2 cm

**Bird bones, including the skull,
of a roller-type bird.**

**Breast bone
of an ibis-type bird.**

Modern tinamous bird.
Wiki Commons
Tony Castro

A small bone from a Tinamous-like bird,
found at Walton-on-the-Naze
by Sue Gregory.

**Fossil bones of a
Phorusphacid 'terror-bird'.**

**Reconstruction of a 2-metre-tall
Phorusphacid 'terror-bird'.**
Wikimedia Commons Nobu Tamura

Fossil Wood & Plants from the London Clay

Rutaspermum seed
Walton-on-the-Naze

Bursericarpum aldwickense seed
Ockendon

Fossil *Nipa* palm fruit
Walton-on-the-Naze

Seeds of *Vitus pygmaea*
Ockendon

Carpolithus pusillus carpel
Ockendon

Platycarya richardsoni fruit
Walton-on-the-Naze

Fossil wood bored by *Teredo* 'shipworms'
Walton-on-the-Naze

Modern *Nipa* palm fruit
(Thailand)

Fossil wood preserved in calcareous nodule
Walton-on-the-Naze

Fossil wood preserved in pyrite
Walton-on-the-Naze

The London Clay also contains one of the world's most varied fruit and seed floras, with over 500 plant species recorded. Of the fruits and seeds found, one of the most common is the fruit of the palm *Nipa*, a plant which survives today only in the mangrove swamps of South-east Asia. Through the study of these fossils a comprehensive picture has been put together of plant life during early Eocene times. Pieces of fossilised wood ranging in size from twigs to large logs are commonly found in the London Clay; they are often riddled with holes, which were bored by the wood-boring bivalve *Teredina*, much like the present day shipworm *Teredo* which bores into wood floating in the sea. Fossilised wood, replaced by the mineral pyrite, is often found washed out of the London Clay onto the foreshore

▲ A river taking driftwood and mud into the London Clay sea – a depiction by Ben Perkins.
Ashley Cooper

around the Essex coast; at Walton-on-the-Naze, for example, parts of the beach are from time to time found to consist almost entirely of well-preserved twigs and small fragments of fossilised wood. All the plant fossils found in the London Clay would have been brought down rivers in large floating masses of vegetation and carried out to sea.

Problems with fossils preserved in pyrite

Many fossils from the London Clay contain pyrite and some are composed entirely of pyrite. 'Pyrite disease' is the main cause of deterioration of these fossils. Pyrite is the mineral iron sulphide which combines with oxygen and water vapour in the air to form, among other things, sulphuric acid and iron sulphate. Sulphuric acid corrodes the specimen and rots labels and storage boxes. Iron sulphate and other salts have a larger volume than the original pyrite and cause the fossil to slowly expand and fall apart. If you leave a pyrite fossil in a card box with a label, it is likely that the box and the label will be destroyed and the fossil itself will turn into a pile of dust.

What can be done? There is often little you can do to prevent pyrite disease. Sometimes the fossil deteriorates very quickly, in days to weeks, other times it lasts much longer. It is difficult to predict. Some specimens survive for years and it seems to depend on the state and size of the pyrite crystals within the specimen. The best that can be done is to isolate the specimen in a sealed container or plastic bag in low humidity. Silica gel granules, such as those that come in packs containing leather goods, can be sealed with the specimen to try to keep it dry. It is important to inspect specimens frequently and discard those that have been affected so that the acid does not spread. Plant material can be particularly difficult to keep and it is best to photograph specimens to retain a record. More sophisticated methods are used by serious collectors and museums, but it is still recognised as a major problem.

▼ A fossil in a collection affected by 'pyrite disease'. Within the specimen salts have formed, expanding and destroying the material; the resulting acid has also attacked the label. *Richard Newton*

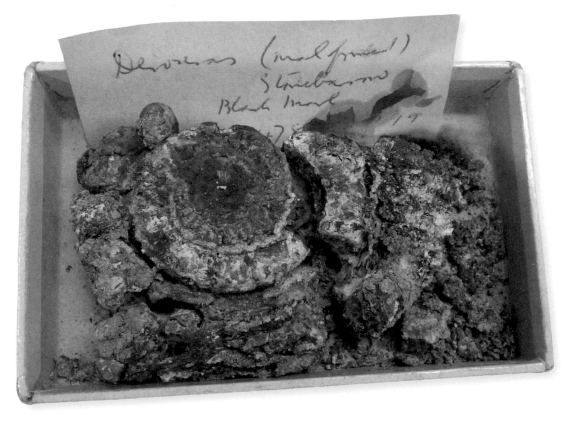

Treasures from the Clay

Some minerals form beautiful crystals which are prized by collectors. Good mineral specimens are rare in the relatively young and soft rocks of Essex, but they do occasionally occur in the London Clay.

Pyrite (iron sulphide), also known as 'fool's gold', forms concretions in the clay and often occurs around pieces of fossil wood.

Baryte (barium sulphate) forms in groups and rosettes of crystals in contraction cracks (septa) in septarian nodules. Sometimes attractive specimens of baryte and green iron-bearing calcite line the septa.

Calcite (calcium carbonate) occurs in green or yellow banded veins, particularly at Harwich and Wrabness. It is very attractive when cut and polished.

Selenite, a variety of gypsum (calcium sulphate) forms clear crystals, usually within or just below the weathered zone of the clay. It results from the chemical reaction between sulphuric acid – formed from oxidising pyrite in the weathered zone – and the calcium carbonate of fossil shells.

◄ Crystal of selenite from the London Clay from the M25 motorway excavations.
David Turner

◄ Radiating crystals of baryte on calcite from a septarian nodule collected from the cliffs at Southend before 1837.
Gerald Lucy

Intergrown pyrite ► crystals (pyritohedral habit) encrusting pyritised wood from the London Clay, Harwich. Field of view 2 cm in width.
Gerald Lucy

▲ Polished slab of banded calcite from the London Clay of Harwich. Length of specimen 11 cm.
Gerald Lucy

Shallow seas to deltas

The Claygate Beds

Towards the top of the London Clay, thicker and more frequent seams of fine sand and silt were laid down as the coarser sediments from the land built up seawards by about 51 million years ago. This change marks the transition to a unit known as the Claygate Beds, but the precise position of the junction is controversial. The Claygate Beds were once known as 'passage beds' because they form the transition between the London Clay and the Bagshot Sand. Detailed correlation between the Essex strata and the type locality at Claygate in Surrey is complex and limited by lack of exposure or borehole information. The 'Essex Claygates' are now considered separately for classification purposes and are from 17 to 25 metres thick over the county.

The boundary of the Claygate Beds with the London Clay is taken at the base of the first fine-grained sandy bed that can easily be distinguished in the field. Springs often form along this line, as it rests on the impermeable London Clay. The boundary probably gets older westwards towards the sediment source. Both the distribution and types of fossils found, together with the rapid and irregular changes in grain size of the sediments, suggest deposition by fluctuating tidal currents. Water depths of less than 30 m are indicated by the fossils, shallowing to less than 10 m in the higher layers. Fossils of the cockle *Venericardia trinobantium* are found only in these beds. There is a pronounced erosion surface seen midway through these deposits when they have been cored in boreholes. This represents a pause in sedimentation during which extensive burrowing of the seabed took place, disrupting any fine bedding detail. This divides the unit into two parts.

1 cm

▲ A small exposure of Claygate Beds in a stream bank on the Pebble Walk Trail at Thorndon Country Park.

◄ Claygate Beds index fossil *Venericardia trinobantium* from M25 excavation, Beredens Lane, Cranham, collected by Graham Ward. *Essex Field Club Collection*

The lower part of the Claygate Beds is seen mainly in west central Essex around Brentwood and on the high ground at Thorndon Country Park. It is a mottled yellow and grey sandy, silty clay which forms a very cohesive land surface. It can be seen in the root plates of fallen trees and in mounds marking boundaries which persist for many decades. There is a small temporary exposure in a stream on the geological trail in Thorndon Park where it can be compared with the much more plastic London Clay close by.

The upper part of the Claygate Beds is mainly thinly interbedded sand and clay with wavy-bedded sands units and very little bioturbation. It occurs at the surface in eastern Essex around Hadleigh and Southend-on-Sea. Regional correlation shows that it fills channels within the burrowed deposits below. These channels may have been caused by the flow of water from the delta advancing from the west.

As the sandy shoreline delta sediments spread further into the sea, they were reworked by offshore currents and deposited as beds of very fine, thinly bedded yellow sand called Bagshot Sand. This contains glauconite, indicating a marine environment, but any organic fossil material such as shells has been leached out. All that remains are occasional moulds of molluscs such as the very long-lived genus *Lingula*, which confirm the marine environment but are not suitable for use in the correlation or dating of these sediments. This makes it difficult to correlate with deposits in a similar stratigraphical position west of London, at Bagshot itself. There, the deposits are coarser grained, non-glauconitic sands and were probably laid down within the upper parts of the delta and not in the sea.

The Bagshot Sand

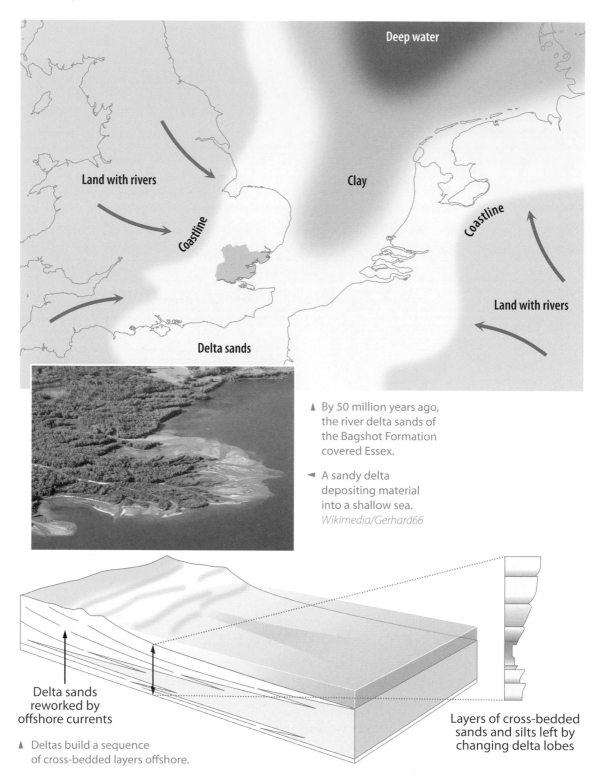

Deep water

Land with rivers

Clay

Coastline

Coastline

Land with rivers

Delta sands

▲ By 50 million years ago, the river delta sands of the Bagshot Formation covered Essex.

◄ A sandy delta depositing material into a shallow sea. *Wikimedia/Gerhard66*

Delta sands reworked by offshore currents

Layers of cross-bedded sands and silts left by changing delta lobes

▲ Deltas build a sequence of cross-bedded layers offshore.

The Bagshot Sand of Essex is very fine-grained, pale yellow to orange-brown. The grains are angular, but very well sorted, resembling 'egg-timer' sand. Some layers show distinctive cross-bedding. It caps the high ground in central and southern Essex at High Beech in Epping Forest, Havering-atte-Bower, Brentwood, Kelvedon Hatch, Billericay, Rayleigh and the Langdon Hills, and forms typically rounded hills. South of Hadleigh, the prominent, rounded hills are capped by Bagshot Sand, such as Sandpit Hill. These are isolated remnants of a continuous deposit that was once up to 25 m (80 ft) thick and must have covered the whole of Essex. Much was removed by erosion when the area was uplifted from around 40 million years ago. It is still being eroded and, in places such as Langdon Hills, sunken 'hollow lanes' have been worn into the sandy hillside. The sandy hills that remain have been protected by the much later, overlying gravel deposits from south-bank tributaries of the ancestral Thames.

A section in Hadleigh Country Park, part of a geological trail, shows layers with ▲ cross-bedding. The section correlates with the lowest part of the Bagshot Sand cored in the BGS Hadleigh borehole nearby. There are very few places in Essex where Bagshot Sand beds can be viewed.

▼ Cross-bedding in a river delta, Bagshot Sand at Hadleigh.

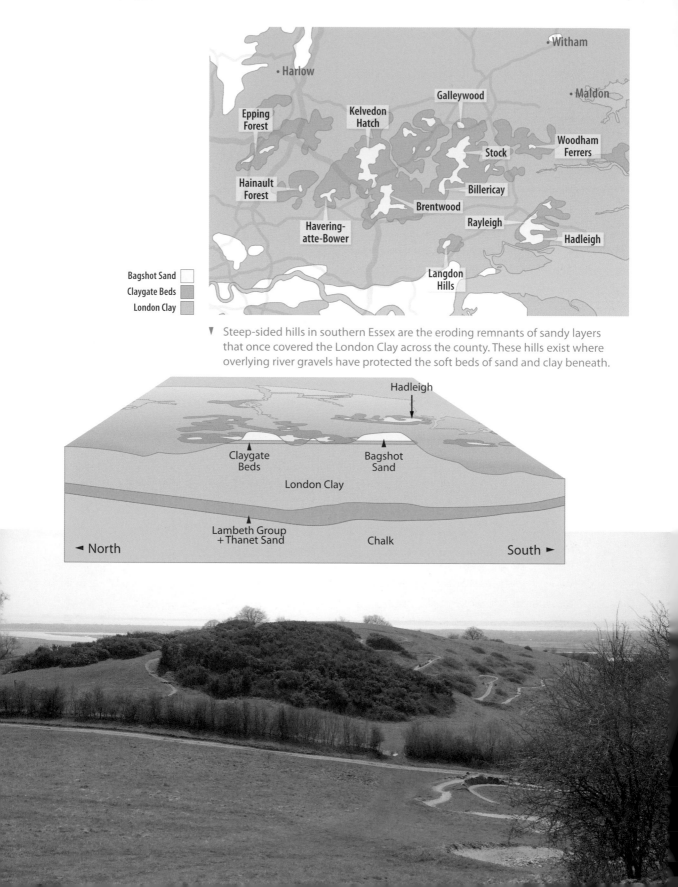

Steep-sided hills in southern Essex are the eroding remnants of sandy layers that once covered the London Clay across the county. These hills exist where overlying river gravels have protected the soft beds of sand and clay beneath.

▲ Landslip on A12 near Colchester in 2013.
BBC News

Many of the landslips around Chelmsford and Brentwood, and in the Langdon Hills and Hadleigh areas of south Essex, are associated with the base of the Claygate Beds, where springs arise from the sandy layers at the junction with the London Clay. Over-steepening is caused by springs undercutting the slopes. Together with the summer drying contraction and winter expansion of the underlying clays, this leads to instability and slipping. Landslipping has also been a problem more recently in large-scale construction sites, particularly along roads, such as the Brentwood bypass and other stretches of the A12.

The London Clay and its succeeding sandy layers demonstrate the impact of Essex bedrock geology on land and on society. The underlying geology and varied nature of the rock layers, however soft or 'crunchy' these are – and however hidden from view – greatly influence the scenery.

Human use of and interference with the land are constantly impacted by the underlying geology. Later chapters feature the importance of geology in Essex for land use and society in greater detail.

◀ One of several round hills capped by Bagshot Sand at Hadleigh Park.

8

A reconstruction of
the extinct giant shark
Carcharocles megalodon
catching small whales.
This huge fish, thought
to have been over
12 m (40 ft) long, lived
in warm seas across
Essex 10 million years
ago. Fossil C. *megalodon*
teeth can exceed
10 cm (4 in) in length
and are occasionally
found in Essex, notably
at Walton-on-the-Naze.
*Painting by
Alberto Gennari,
reproduced with
the permission of
Prof. Giovanni Bianucci*

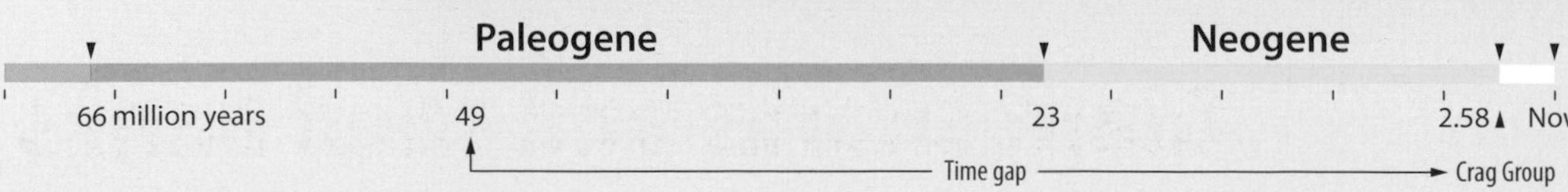

	Paleogene		Neogene	

66 million years   49   23   2.58 ▲   Now

Time gap ————————————————————► Crag Group

Giant sharks and shell banks

A time gap and a cooling climate

During much of the past 50 million years Essex has been land. For this reason, few rock layers were laid down here through this time span. We know from geological exploration outside Essex, such as beneath the adjacent North Sea, that periods of erosion removed rock from across the land. For much of this interval the North Sea area was occupied by a huge delta where the sediment was spread far out towards the north. Global studies reveal that a cooling climate, with large temperature swings, culminated in the present ice age. Furthermore, Essex has been carried almost 10 degrees of latitude further north by continental drift through the last 50 million years, further influencing its climate. All these changes have greatly affected the landscape. Sand layers in the north of Essex show that the sea spread back across the area within the past 3 million years. These layers, together with the subsequent ice age sediments, are the most recent geological deposits in the county.

A large time gap in Essex rock

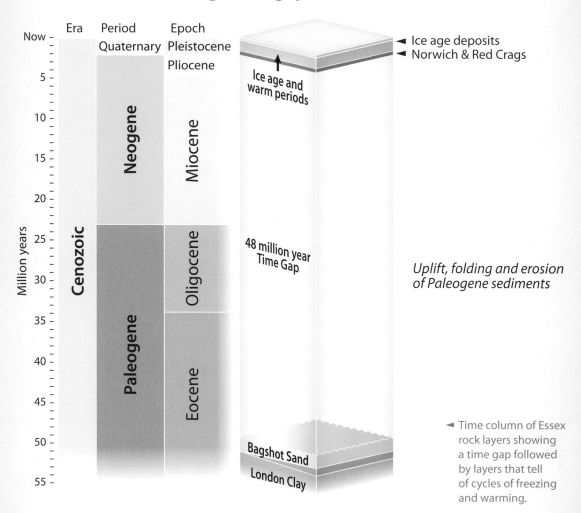

Red Crag

Time Gap

London Clay

Sediments in the North Sea and mainland Europe, as well as rocks and ocean beds across the world, all provide clues to tectonic movements and changes in climate that affected our region through the 48-million-year time gap. The various different pieces of evidence point to a story of squeezing and cooling across Essex.

Following the Bagshot Sand deltas that spread across the London Clay sea 50 million years ago, some sediment may have been laid down across Essex to form rock layers prior to the Red Crag deposits of 2.6 million years ago. However, no such layers remain within the county. Any Essex rocks from these times have been removed by erosion and transported away. Also, because the Essex area was above sea level for much of this time, it is unlikely that extensive, thick

▲ The position of the time gap between the London Clay and the Red Crag seen in the cliffs at Walton-on-the-Naze.

sediments were ever laid down. Episodes of strong tectonic movement of the Earth's crust created the Alps and Pyrenees and this also tilted and compressed south-east England. These crust movements pushed up the rock layers along the southern margin of Essex forming the London Basin; the squeezing also helped to keep Essex above sea level, enabling erosion to eat into its landscape. Much of the Bagshot Sand has been eroded and, in north-eastern Essex, most of the London Clay is missing so that the Red Crag here rests directly on the lowest beds of the London Clay, adding to evidence for this time gap.

Tectonic movements have gradually cracked the thick crust of the Anglo-Brabant Massif beneath Essex and surrounding areas, affecting the rocks draped across the Massif and modifying the landscape. This has influenced how and where new rock layers formed. More recently, the effects of climate cooling, culminating in regular cycles of deep freeze and sudden heating, have amplified the effects of vertical movements in the crust beneath Essex and also brought large changes in sea level. As sediments have eroded away, the land has risen, rather in the way that an iceberg rises under the influence of melting above the sea. Meanwhile, the crust beneath the North Sea received extra sediment brought from the eroding lands, pushing the crust downwards as the sediment load built up. The result is that Essex continues to tilt slowly down to the south-east; as it does so it has once more become an edgeland with seas and marshland advancing and retreating.

At some stage during all these slight movements in the crust, the River Thames turned from its former north-west to south-east flow, into a course across north Essex and through Suffolk. After passing through the Goring Gap in the uplifted Chiltern Hills, the river now turned left, to the north-east. Its new course may have been initiated by slight changes in the landscape caused by downward south-west–north-east rift movements deep in the crust beneath the London Basin. Similar movements appear to have controlled the deposition of the Crag Group sediments further to the north-east.

The closer we come to the present day, the more we can discern from the rock record; and we also realise that there are many smaller time gaps within rock layers. It is rather like peering through a microscope and observing at different levels of magnification. As we travel on a journey through time and rocks, we may observe more recent geological events in greater detail and on much shorter timescales.

▼ Stages of gradual cooling into the current ice age. Antarctic ice developed before 30 million years ago, while periods of global warming have given way to further cooling. Arctic ice has developed as the ice age intensified through the past few million years.

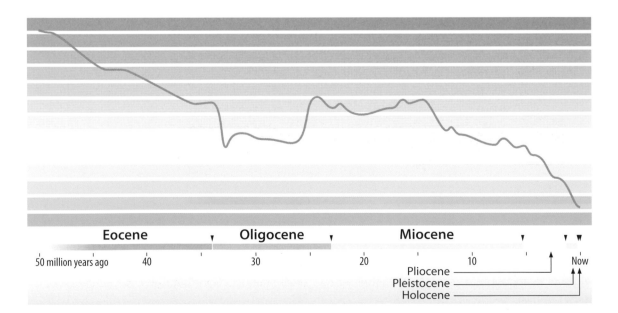

Clues from the time gap: 'coprolites' and boxstones

▲ Research at the junction of the Red Crag upon the London Clay in the Naze cliff.

◄ Phosphate nodules – 'coprolites' – from the junction bed, seen at the lower right within the sieve.

Clues to what happened during the time gap include records of borehole cores from North Sea exploration, and the few remaining pieces of geological evidence in and around Essex. One strange set of clues is contained in the cliffs at Walton-on-the-Naze, within a layer at the junction of the London Clay and the overlying and much more recent Red Crag sand. In the red sand, just above the eroded surface of the grey London Clay, there are abundant dark brown, rounded and shiny nodules popularly known

▲ A broken solid 'boxstone' with small shells, Walton-on-the-Naze. Boxstones offer the sole remaining evidence of any Miocene rock layer in Essex.

as 'coprolites'. These are rich in phosphate and were collected on a large scale in the past, becoming the basis of a fertiliser industry centred on Wrabness and Walton. The name 'coprolite' refers to fossil excreta from various animals; however, these are not excreta. The phosphate, with calcium and fluoride, is derived from the recrystallisation of calcium phosphate-carbonate biomaterials that made up the bones, scales and teeth of marine creatures that lived in or on, or drifted down to, the muddy sea floor. The nodules gradually formed as concretions within the stagnant mud and phosphate mixture.

Many phosphate nodules were formed within the London Clay across Essex. They also occurred in later seabeds, including those that must have existed across parts of Essex during the time gap. Here, in the cliffs and on the beach, is evidence for these past seabed deposits that were laid down after sediments had been eroded by the waves and swept off the former shorelines. The mud and sands were winnowed by the scouring tidal water currents to leave a 'lag' of coarser particles, notably the many dark 'coprolites' scattered in this junction layer.

Even more remarkable, in this junction layer, there are a few well-rounded lumps of very hard sandstone called 'boxstones'. Occasionally, larger fossils are discovered, including bones of land mammals. Some of these fossils indicate that the seas eroded into rocks of Miocene age, and similarly sorted out larger pieces rock and various fossils from that deposit.

◄ A whalebone from the Naze beach, showing its characteristic 'polished' appearance.

A Miocene fossil ► bivalve, a natural internal mould from a boxstone, found at the Naze.
Gerald Lucy

▲ A rare Miocene 'beaked whale' snout bone fossil found on the Naze beach.
Alison Mercer/Phil Chatfield

Those Miocene rocks no longer occur anywhere across the land surface of Essex, having been totally eroded away. Indeed, only very few, tiny deposits of this age exist anywhere in Britain. Here, in this junction bed, are clues to the former existence of those lost rock layers from before 5 million years ago. Their age is indicated by comparing the fossil shells that are found in the boxstones with fossils found in much thicker beds of Miocene age across the North Sea in Europe. This proves that these stones are older than the Red Crag seabed layer in which they are found.

After being eroded out of the base of the Red Crag, the boxstones survive well enough to be found along the beaches of north Essex and Suffolk. Boxstones are made of sandstone cemented with phosphate. A small percentage of the stones contain a fossil hidden in a cavity within. If the stone is hit sharply, it might reveal a large bivalve shell, a shark tooth or even a piece of whalebone; however, most do not hide any fossils and are just solid sandstone throughout, perhaps with a few small shells or fragments embedded.

▲ A whale tympanum (ear bone)
from Walton-on-the-Naze.
Michael Daniels

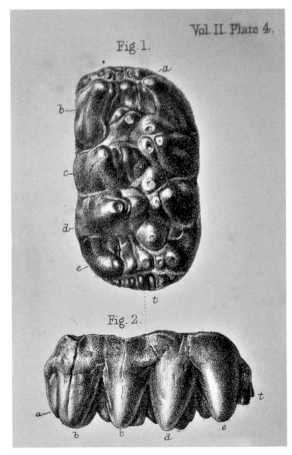

▲ An engraving of the side and top view
of the Ramsey mastodon tooth
in the **Quarterly** *Journal of the Geological Society*, **1857**.
The Geological Society of London

The Red Crag junction bed notably yields fossil teeth of the giant shark *Carcharocles megalodon*, together with rolled and highly polished bones from whales including the ear-shaped tympanum, a whale ear bone. At Walton, digging in the cliffs is not permitted, but fossils from the junction bed can be found on the beach, especially after major cliff falls. Tide tables should be consulted to ensure your safety as high tides can reach up to the cliffs and there is a real danger that you could become part of the material swept offshore and incorporated into a new geological deposit. There is a remarkable collection of specimens in Ipswich Museum from Red Crag sites in Suffolk which provides an insight into the contents of this important junction bed. The high polish shown by many of these fossil bones and teeth appears to be due to a coating of phosphate within the junction bed. However, once rolled by the waves on a modern beach they soon become dulled.

The discovery of an exceptionally fine molar tooth of a mastodon from the parish of Ramsey was recorded in a paper in the *Quarterly Journal of the Geological Society* in 1857 by H. Falconer. He gave a detailed description of the tooth, stating that it was 'lately discovered by the Rev. Mr. Marsden in the bed of coprolitic or phosphatic nodules in the parish of Ramsey in Essex … and kindly lent to us for description'. He stated that it was 4.9 inches in length and up to 2.9 inches wide, and from the upper jaw of the creature, right hand side. Mastodons were relatives of mammoths and modern elephants that became extinct in Britain over a million years ago. The paper contained a fine engraving of the specimen. The current whereabouts of the specimen is not known.

▲ *Carcharocles megalodon* fossil
tooth from the Red Crag
at Walton-on-the-Naze.
Jim Greenwood

Fossil sharks

Many of the rocks of Essex contain large numbers of
shark teeth. Could this be evidence that prehistoric
seas were once teeming with sharks? Teeth are usually
the only part of a shark to be fossilised because the
skeleton is entirely composed of cartilage rather than
bone. This is one of the features that distinguish sharks
from other fish. Another feature is the arrangement
of teeth which exist in rows with only the front row
in use at any one time. The teeth are only loosely
attached to the jaws and are often torn away allowing
fresh teeth to move forward to take their place. This
rapid turnover of teeth is likely to account for the
vast numbers that accumulate on the sea floor to be
preserved as fossils.

▲ The frighteningly large gape of the giant shark
Carcharocles megalodon. Lifesize replica, Ocean Park,
Hong Kong.

The largest of all sharks was the awesome
Carcharocles megalodon whose fossil teeth can exceed
10 centimetres (4 inches) in length. This animal, which
lived in late Miocene times around 10 to 7 million
years ago, is thought to have been over 12 m (40 ft)
long, with gaping jaws that would have been large
enough to accommodate a human. The teeth of this
extinct shark are rare but they can occasionally be
found on the beach at Walton, derived from the Red
Crag junction bed. A technically researched picture
of *C. megalodon* is shown at the start of this chapter.

Crag seas and shell banks

After much of the compression from Alpine movement had ceased, occasional underlying adjustments in the crust have continued to influence sediment accumulation over Essex. Deep faults, some with south-west to north-east trends, plus changing coastlines at the margins of the sinking southern North Sea basin have affected the pattern of geology in the last 3 million years. During this time, at the very end of the Pliocene Epoch, Essex was covered by a shallow sea, up to 15 to 25 m (50 to 80 ft) deep. The climate was variable but gradually cooling; the sea advanced and retreated with the tilting of the crust and with changes in climate. During the sea's advances, beds of sand, some containing an abundance of marine shells, were laid down on the seabed in the succession of layers that form the Crag Group. The word 'crag' was formerly a local term used in East Anglia to describe shelly sand, but geologists have now adopted it to designate these marine deposits of Pliocene to Pleistocene age in Essex, Suffolk and Norfolk, collectively termed the Crag Group.

The oldest formation of the Crag Group is the Coralline Crag, found just to the north of Essex where it forms a buried ridge between Aldeburgh and Orford in Suffolk, extending south-west to the River Deben and to Tattingstone, south of Ipswich. Its relationship to the overlying Red Crag Formation shows that this ridge formed islands in the later Red Crag Sea. The Red Crag Formation underlies areas of north Essex from the coast and inland to around Stansted Mountfitchet and into Hertfordshire. Overlying the Red Crag in north Essex is the slightly less extensive Chillesford Sand, taken to be part of the Norwich Crag Formation.

▼ The Red Crag sea across Essex and East Anglia.

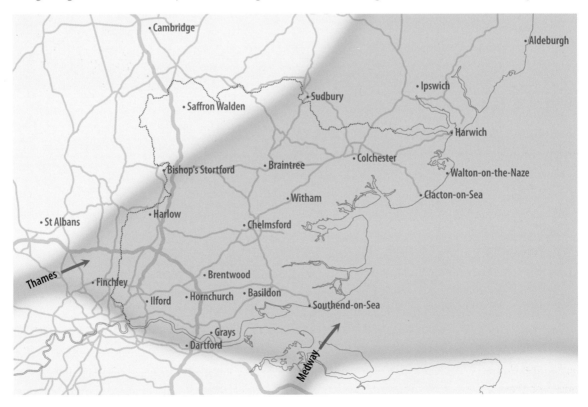

▲ Life in the Crag Sea.
Louis Wood/Roger Dixon

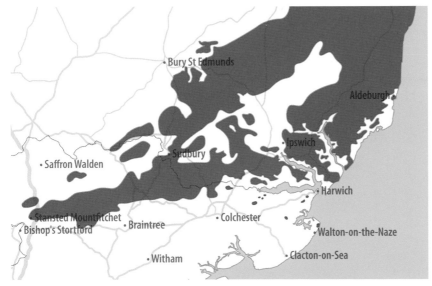

◄ The occurrence of Crag Group deposits across Suffolk and north Essex. Most of the Essex Crag is obscured by later superficial layers and only a few small outcrops appear, most notably at Walton-on-the-Naze where Red Crag is seen in the cliffs.

■ Crag deposits including Chillesford Sand

The ages of the Crag formations are based on their stratigraphical relationships together with biostratigraphy based largely on pollen and other microfossils. However, these are as much climate dependent as age related, so clear absolute age determination is difficult. The Red Crag at Walton is older than the Red Crag of Suffolk and was laid down more than 2.6 million years ago.

Red Crag in Essex

The most spectacular exposure of Red Crag in East Anglia is at Walton-on-the-Naze where, in the cliffs to the north of the town at the Naze itself, the red, shelly sand can be seen resting on the blue-grey London Clay with the junction bed containing phosphate nodules and phosphate-rich fossils described above. Walton is one of the finest geological sites in Britain; both the Red Crag and the London Clay here are of international importance and the cliffs are designated a Site of Special Scientific Interest (SSSI).

In Essex, Red Crag was formerly thought to exist only in the Walton area and in other isolated patches in the north-east of the county, but surveys carried out by the British Geological Survey in the 1980s revealed that it extended west as far as Stansted Mountfitchet, although it is mostly buried beneath

▲ Red Crag in the Naze cliff, a geological Site of Special Scientific Interest (SSSI) with a close-up of the shells in the Crag beds. This site is becoming obscured by vegetation now that the coastal defence of the Crag Walk is protecting the cliff from rapid collapse.

later superficial deposits. In the west of Essex, it had not been previously recognised because much of the sand has been decalcified, that is, the fossil shells have been dissolved by percolating groundwater. Red Crag has been exposed in gravel pits such as at Canfield, Elsenham, Halstead, Stebbing and Widdington. The range of elevations of the base of this bed from 130 m in Hertfordshire to 90 m at Stansted and less than 10 m at the coast shows that considerable crustal tilting was still occurring after its deposition, with the land rising to the north-west coupled with subsidence of the North Sea basin, where much thicker layers of Crag accumulated.

Inland borehole cores show the Red Crag sands to be coloured green at depth, due to the mineral glauconite which forms in shallow marine waters. The rust-red colour of the sand in the Walton cliffs and wherever else the deposit is at or very near the surface is due to the reaction of the green glauconite grains with oxygen in air and groundwater. This effect is perhaps aided by the oxidation of pyrite eroded from the London Clay into the overlying Red Crag. The final product of these chemical changes is bright red iron oxide and hydroxides which stained the sand grains and fossils. The local occurrence of 'iron pan' where groundwater has concentrated these iron compounds, often in thin seams along bedding planes, is also characteristic and may help to establish the beds as Red Crag when fossil shells are absent.

The Red Crag at Walton displays sedimentary structures which are evidence of the conditions that existed on the seabed at the time of deposition. Of these, 'cross-bedding' is the most common, where the movement of seabed sand banks by strong currents has formed inclined layers like underwater dunes; from this evidence it is possible to establish current

strength and direction. It is likely that each sand bed was laid down over a very short time, perhaps in one storm event, and the whole sequence may have been deposited within only decades. Red Crag deposits further north in Suffolk are probably from later sea incursions over the course of half a million years, followed by later layers of Crag Group sediments. Some fossils at Walton-on-the-Naze indicate a warm climate; however, these shell banks were laid down during a time when the climate had started to deteriorate, with alternating cold and warm periods each lasting tens of thousands of years.

Red Crag at the base of Elsenham Quarry, ▶
north-west Essex. *Gerald Lucy*

▲ Tidal cross-bedding in the Red Crag cliff at the Naze.

▲ A burrow found on the Naze beach.

◄ Burrows filled with sediment in the Red Crag at the Naze.
Gerald Lucy

The Crag layers contain cylinders of cemented sand which are the filled-in burrows of marine animals; these are known as trace fossils and are valuable clues to the movement of burrowing animals and other life on and in the sea floor. Some of these burrow fillings are harder than the sand around them and these are eroded out and can be seen on the beach. They often show concentric bands of colour in cross-section. They vary in size: large ones up to 10 cm in diameter were probably formed by lobsters; smaller ones may be worm burrows.

Small isolated patches, or outliers, of Red Crag occur in north-east Essex. One such patch caps the top of the hill occupied by Beaumont Hall, a little to the west of the backwaters behind Walton-on-the-Naze. It is one of the few fragments of the once-continuous deposit of Red Crag across north Essex that has now been diminished by erosion.

The Red Crag at Beaumont was first brought to the attention of the scientific world by the well-known

▲ The view east from the hill at Beaumont across to the backwaters north of the Naze.

Essex geologist John Brown of Stanway, who obtained over 90 species of fossil shells from a pit near the south-eastern edge of the outlier. He privately published a list of these in 1846. At the end of the nineteenth century the amateur geologist Frederick Harmer (1835–1923) carried out a detailed study of the fossils of the Red Crag. Harmer was a Norwich wool merchant and textile manufacturer who was also Mayor of Norwich between 1887 and 1888. In his spare time, however, he was a specialist in fossil molluscs and his thorough study of British Tertiary and Quaternary deposits provided the basis for today's accepted views. He re-opened Brown's pit, where he succeeded in finding many more species. Most of these were characteristic of the Walton Red Crag but a few were northern cool-water species which were rare or absent at Walton, indicating that this outlier may be slightly different in age from the Walton Crag.

With the permission of the landowner, Harmer also dug a hole near the south-western limit of the Beaumont outlier which revealed 5 or 6 feet of Red Crag resting on London Clay. With the help of a labourer 7 or 8 tons of Crag sand were sifted and from this several species were encountered that were not found at the previous pit. Harmer published a detailed account of his work at Beaumont and elsewhere in the *Quarterly Journal of the Geological Society* in 1900. In that year the mineral content of the Red Crag at Beaumont was analysed by treating a 25 g sample with concentrated hydrochloric acid. What remained was 50% quartz, but the other 50% contained a rich assemblage of other minerals such as feldspar and mica and rarer minerals such as zircon, rutile, kyanite and ilmenite, together with yellow crystals of sapphire and green and blue grains of tourmaline. There were also tiny red garnets that were so plentiful that the heavy concentrate had a strong pinkish colour. Grains of topaz have also been found in the Red Crag although none was recorded from the Beaumont sample. The presence of such a suite of minerals in the Red Crag is evidence that these sands were partly derived from metamorphic rocks, possibly from the erosion of the Scottish mountains, and may have been

▲ Harmer's excavation in the Waltonian Red Crag at Little Oakley.
From the Frederic William Harmer collection, British Geological Survey © UKRI 2021

recycled several times over hundreds of millions of years before being deposited on the floor of the Red Crag Sea. As the Red Crag continues to be eroded, these minerals will end up on the floor of the North Sea – the next stage in the continuous 'rock cycle'. The former Red Crag exposures at Beaumont are no longer accessible, but fossil shells can be seen scattered on arable fields, such as by the public footpath east of Beaumonthall Wood. Red Crag sand and shells have also been seen around animal burrows in the ditch bank alongside the footpath west of Beaumont Hall.

At Little Oakley, to the north-east of Beaumont, the existence of shelly Red Crag capping the high ground has been known since at least the 1860s but it was not until the work of Harmer that the site received any attention. The invention of the motor car gave Harmer the opportunity to do more geological field work. With the permission of the landowner, he reopened a shallow pit to the east of the village near Foulton Hall and sieved the sand for fossil shells over a period of several years.

The result of these efforts was his two-volume work *The Pliocene Mollusca of Great Britain* published in 1919 in which he states, rather enthusiastically, that over 600 different species of mollusc were found in this pit (nearly 400 of the species illustrated in the book are from Little Oakley). Harmer's work has shown the extraordinarily rich molluscan fauna of the Red Crag sea. Harmer records that all of the fossils came from 'an area of twenty yards square' and says that they were obtained 'during many years labour, and by the sifting and examination of something like 200 tons of material'. Harmer's pit has long ago been filled in but between 1973 and 1975 the Ipswich Geological Group carried out an excavation near Foulton Hall to re-expose the fossiliferous Red Crag and found a large number of fossil molluscs. The 1976 report contains a detailed faunal list and an illustration of the section.

More recently, between 2002 and 2004, two separate excavations were carried out near Foulton Hall by the late amateur collector John Hesketh. His collection of molluscs from these sites has been donated to the Essex Field Club.

There was an exposure of shelly Red Crag on the coast at Harwich, but this was entirely destroyed by the action of the sea in the nineteenth century. It was, however, recorded by Samuel Dale in 1730 in his book, *The History and Antiquities of Harwich and Dovercourt*, which was the first book to describe and illustrate fossils from Essex.

▼ A cliff of Red Crag upon London Clay that formerly existed at Harwich. From *The History and Antiquities of Harwich and Dovercourt by Samuel Dale, 1730. Reproduced by courtesy of the Essex Record Office*

Tab. VIII. *Page 99.*

A. Land-guard-Fort. B. *The Andrews or Barr of Sand runing from the Fort.* C. *The South or Dover Court Point of the Havens mouth.* D.D. *The Strata of* Sand, Gravel, Fossil-Shells, *and their Fragments.* E. *The Stratum of* Blewish Clay *divers feet deep.* F. *The heep of* Gravel, Sand, & Shells, &c. *which caveing down from the Top ly at the bottom of the Cliff.* G. Cliff-Stones. *which ly upon the Shore before the Cliff.* H. *Persons observing The Cliff.* I. A *Stone of the Lower or Stoney Stratum.*

Red Crag fossils at the Naze

The Red Crag at Walton is highly fossiliferous and some layers consist almost entirely of fossil shells, many of which are broken fragments. Some whole, single bivalve shells are distributed by currents along bedding planes. The fauna is diverse and well preserved and hundreds of species have been recorded. However, the fossils are often fragile. Corals, echinoids and barnacles can be found, but by far the most common fossils are bivalves such as the dog cockle *Glycymeris variabilis* and gastropods such as the 'left-handed' whelk popularly known as *Neptunea contraria* but now renamed *Neptunea angulata* due to the use of its previous name for a modern whelk found off the Portuguese coast. This whelk is 'contrary' as it spirals the 'wrong way', in other words in the opposite direction to almost every other gastropod. Shark teeth can be found within the Red Crag, most being derived from the underlying London Clay, but some are from Miocene and Red Crag sharks. Because the

Red Crag layer is crumbly and the cliffs are so unstable, the supply of fossils onto the beach is occasionally replenished, particularly after winter storms and high spring tides. The cliffs themselves are designated as an SSSI and should not be dug into and they are also potentially dangerous; however, during a falling tide the beach is an excellent discovery area for fossils from the Red Crag as well as from the London Clay, and also

▲ A young geologist's finds on the Naze beach.

▼ Red Crag fossils in shelly sand debris at the top of the Naze beach.

▲ *Neptunea angulata* (usually known by its older, more descriptive name *Neptunea contraria*) the 'left-handed whelk'. A right-handed modern whelk shows the more usual helical whorl.

▲ Official research and cleaning at the Naze cliff Red Crag SSSI in 2017 – vegetation and rain-wash gradually obscure the outcrop. The Crag here is overlain by clay and silt layers and then the sands and gravels of the later ice age Thames and Medway riverbeds.

▲ Shark's tooth within the shelly Red Crag, Naze cliff SSSI.

for interesting pebbles from the overlying riverbed gravels plus pieces of 'copperas' and 'coprolites'.

Red Crag fossil shells are mostly stained reddish-brown due to the iron oxide in the crag beds. There are useful displays of fossils and other beach finds in the Naze Tower on the clifftop and also in a cabinet assembled by GeoEssex in the nearby visitor centre run by the Essex Wildlife Trust. Here you can check the names and compare your collection from the shore.

As we continue with our story of the geological history of Essex and encounter these younger Crag rocks, more modern animals and plants are found in the fossil record. It has been estimated that about 3% of species living 50 million years ago still exist and that 50% of species living 3 million years ago exist today.

◄ Living Moon Snail from East Timor and fossil relative *Natica crassa* from the Red Crag. *Wikimedia Commons /Nick Hobgood*

Fossils from the Red Crag

Cup coral
Sphenotrochus intermedius

Necklace shell
Natica crassa

Rock whelk
Nassarius recticosus

Rugged dog whelk
Spinucella tetragona

Hydrozoan
Hydractinia

Venus shell
Venus casina

Rayed artemis
Dosinia exoleta

Dale's whelk
Leiomesus dalei

Sea urchin
Echinocyamus pusillus

Cowrie shell
Trivia coccinelloides

Dog cockle
Glycimeris variabilis,
formerly
Glycimeris glycimeris

Giant cockle
Cerastoderma parkinsoni

Smooth whelk
Colus curtus

Curved moon shell
Euspira catenoides

Tower shell
Turritella incrassata

Astarte shell
Astarte obliquata

Trough shell
Spisula arcuata

Queen scallop
Aequipecten opercularis

Spiral sea snail
Euroscaphella lamberti

Left-handed whelk
Neptunea angulata,
formerly
Neptunea contraria

Chillesford Sand and River Thames

Across the north of Essex, and lying directly on the top of the Red Crag, is a layer of sand up to 10 m (30 ft) thick, the Chillesford Sand (Chillesford is in Suffolk). It was laid down in a shallow sea embayment and estuary around 2 million years ago, some half a million years after the Red Crag at The Naze. At this time, the inlet of the sea reached into west Essex where it was probably part of the ancient Thames estuary, receiving much of its sand and mud from that river. The Chillesford Sand varies in colour from pale yellow and orange to grey with occasional seams of flint pebbles. The change in colour from the more intense orange-red of the Red Crag probably reflects a change in the climate and perhaps a more estuarine location.

The Crag Group is usually considered to be part of bedrock geology according to the British Geological Survey. Chillesford Sand is classed as part of the Norwich Crag, within the Crag Group. However, based on historical surveys, Chillesford Sand is considered to be a superficial deposit, and is shown as such in maps and publications, whereas the Red Crag of Suffolk and Essex is considered to be bedrock. The whole of the Crag Group was deposited after a large time gap and represents a set of sediments extending from the present mainland across the North Sea area during the gradual onset of the ice age. Because of this, the whole of the Crag Group – including the Essex area Chillesford Sand – is shown simply as a single superficial deposit in geological maps within this book. The Thames gravels that fed into these various deposits are all considered to be superficial, so this lends further weight to our unification of the whole Crag Group as superficial deposits. Furthermore, sparse Chillesford Sand fossil evidence in Suffolk indicates that it might be distinct from the rest of the Norwich Crag and deposited as a superficial layer from ice age estuary outflow, including that of the Thames.

▼ Chillesford Sand overlying Red Crag in a quarry at Elsenham in north-west Essex. *Gerald Lucy*

Although fossils occur in the Chillesford Sand in Suffolk, none has been found in this deposit in Essex. Complex sedimentary structures in fine- to medium-grained sandy beds, usually less than 0.3 m thick, consisting of tabular and trough cross-bedding and ripple lamination, indicate the action of tidal currents. Quieter conditions are indicated by the interbeds of laminated clays. The absence of fossils and the abundance of water escape structures appear to indicate rapid sediment accumulation. The Chillesford Sands are rarely seen at the surface in Essex as they are largely buried beneath later ice age superficial deposits. Both the Red Crag and the Chillesford Sand have been quarried, as at Elsenham in north-west Essex.

The pollen and microfossil content of the beds of sand, silt, clay and gravel above the Red Crag towards the top of the cliff at Walton-on-the-Naze reveal climate changes leading into the ice age. The sand and clay beneath the Thames-Medway gravels could be from meandering rivers such as the Medway into an estuary on the North Sea coast, at any time from that of the Chillesford Sand about 2 million years ago, to that of the succeeding Thames-Medway gravel layer of around 500,000 years ago.

With the continued tilting of the region to the south-east, the seas retreated and rivers spread their sand and gravel along courses that varied with the underlying crust movements. Meanwhile the climate was swinging between warm and ever-colder as the

▼ Chillesford Sand estuary 2 million years ago, with possible river courses.

ice age approached, creating repeated changes in the course and flow of the River Thames and the position of the Essex coastline – the ever-changing edgeland. As the ice age deepened, it triggered huge cycles of change. Eventually, most of the land surface of the county was covered with layers of new superficial sediments, drastically altering the landscape to the one in which we now live. Tectonic shifting and tilting and swings in climate continue to change the landscape and the positions of coastlines and rivers across Essex to this day.

▼ Layers above the Red Crag at Walton-on-the-Naze reveal the onset of the ice age.

Soil and ice age wind-blown silt

Ice age Thames Medway river gravel

River clay, silt and sand layers

Red Crag marine sand

9

Summertime in Essex during the Anglian glaciation 450,000 years ago. The edge of the ice sheet across northern Essex lies on the distant horizon. Our steppe mammoth *Mammuthus trogontherii* is 4 metres tall at the shoulder and weighs 10 tonnes. The ancient Thames flows past, across mid-Essex where Chelmsford is today – a huge braided river in a time of snowmelt. *Painting by Peter David Scott*

Neogene ▾	Quaternary	▾

3 million years 2.58 ▲ 2 1 Now
Onset of hot and cold stages Cold intensifies ▲ Ice sheet over Essex Anglian ▲▲ Purfleet hunters ▲

Ice age Essex

Time and change

Essex is on the front line of big changes in climate and sea level. For much of its history the Earth's surface has been warm and free of ice but, occasionally, global temperatures drop and the world experiences an 'ice age'. During each downturn in climate, lasting tens of millions of years, areas of ice wax and wane across polar regions and mountain ranges. We are all living in the Quaternary ice age right now. Large spreads of ice have formed around the Arctic and the Antarctic. Through this time, the planet has become cooler on average than at any time since the previous ice age more than 250 million years ago.

Nevertheless, within this overall cooling there have been dramatic swings in the climate. The impact on Essex has been – and continues to be – severe.

▼ Five million years of cooling. The start of the ice age is drawn at 2,580,000 years ago, after which the climate has suffered increasingly profound heating and cooling events. The greatest extent of ice during a cold swing – a glaciation – was during the Anglian Stage 450,000 years ago. There have been four further glacial stages since then. The last one ended only 11,700 years ago.

Much of human evolution took place during the Quaternary Period. Its start defines the beginning of this current ice age. The Quaternary is subdivided into two Epochs: the Pleistocene takes us from 2,580,000 years ago to 11,700 years ago; the Holocene is the time from 11,700 years before the year 2000 and marks the current warm interglacial of the ice age continuing to the present day. Thus, the Holocene Epoch is merely the short warm stage following the latest violent swing in climate.

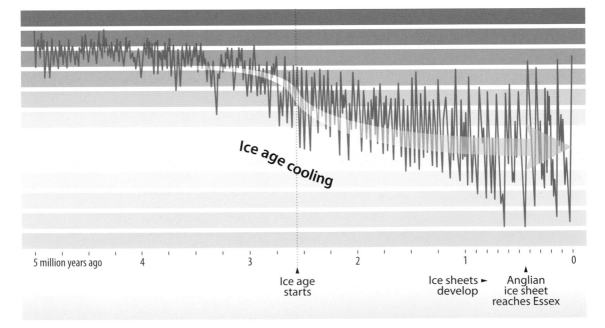

Ice age cooling

5 million years ago   4   3   2   1   0

Ice age starts   Ice sheets ► develop   Anglian ice sheet reaches Essex

Essex rock layers and ice age time

The surface layers of Essex are a geological product of the many climate swings through the last 2.5 million years. The overall effects of these changes are seen in three major episodes through this time span. They have left their evidence in the county's rocks, soils and landscape. The diagram shows the rocks of Essex in a time sequence.

From about 2.5 million years ago, the seabed sands of the Red Crag and Chillesford Sands were laid down across north Essex; a series of gravel riverbeds of the ancient River Thames spread across much of north and west Essex. Then, less than half a million years ago, ice extended south across Britain as far as London, blocking and diverting the Thames. During this short interval (shown in blue in this diagram), much of Essex lay beneath the Anglian ice sheet, which left behind a thick blanket of ground-up rock. Following diversion by this ice, the post-Anglian Thames again spread its sand and gravel, but this time across southern and eastern Essex.

The story is told by these layers of Essex rock. They provide evidence of repeated swings in climate, of big rivers and deep permafrost. A remarkably complete ice age story has been prised out of the land from hand-dug pits, big quarries, wells and boreholes, from cliffs and the soil.

Observers have gradually pieced together a long and eventful ice age county history. This is an unfinished account of ice, wind and water and the comings and goings of humans. This detailed story can only increase our appreciation of the landscape – and the nature of present-day changes.

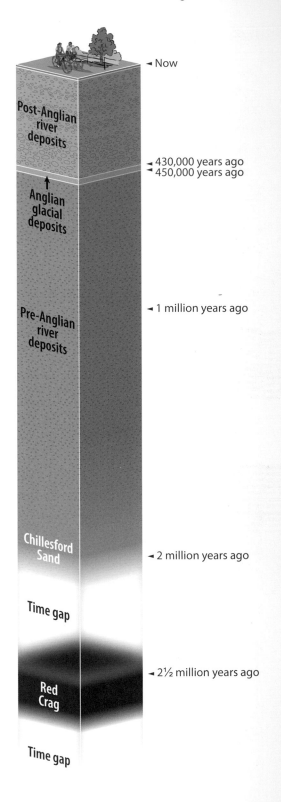

A time sequence for ice age Essex. The larger ► time gaps between rock layers are shown. There are many smaller time gaps (not shown).

Ice age Essex in three episodes

1. Pre-Anglian: the early Thames riverbed layers across north-west and mid-Essex, known as the Kesgrave Sand and Gravel, predate the spread of the Anglian ice sheet across the area.

◄ Pale sandy gravel of the pre-Anglian Thames riverbed, Bull's Lodge Quarry, Boreham.

2. Anglian: glacial till in the Essex area, the Lowestoft Till or 'chalky boulder clay', was spread across a large part of the county in the base of the Anglian ice sheet.

◄ Chalky till from the Anglian ice sheet 450,000 years ago, Bull's Lodge Quarry, Boreham.

3. Post-Anglian: sands and gravels of the diverted Thames occur in southern and eastern Essex together with significant interglacial layers.

◄ Quarrying post-Anglian Thames gravel at Fairlop.

Formal names used in the geological literature for ice age geological deposits across Essex, with some general and informal names.

Unit defined by British Geological Survey (BGS)	Sub-unit or description	General and informal terms
	Periglacial and post-glacial deposits	Head, brickearth, loess, 'hoggin', lake bed silts, etc.
Thames Catchment Subgroup	Maidenhead Formation (Post-Anglian Thames terraces)	Terrace gravels, soils; brickearths and other interglacial layers
Lowestoft Formation	Anglian glacial till: outwash sands and gravels; lake deposits	Till; chalky boulder clay: glacial outwash; lake bed silts
Kesgrave Catchment Subgroup	Colchester Formation	Kesgrave Sand and Gravel; soils and interglacial layers
	Sudbury Formation	

▼ This geological map across Essex reveals three episodes of the ice age. It shows the three layers of sediment left before, during and after the great spread of Anglian ice across a large part of Essex. These soft rock layers are being cut into by later erosion to leave a tattered picture of river valleys, terraces, farmland plateau – and a few steep hills. Most of the Anglian glacial deposits consist of chalky till ('boulder clay'); the river deposits are mostly of sand and gravel.

Why are we in an ice age?

Global cooling has multiple and complex causes. Long-term alterations in deep ocean circulation as continents drift together and apart and seaways open and close, the gradual uplift and erosion of mountain chains – all of these greatly affect world climate. The current ice age has been a gradual process that, across the area of Essex, started around 2.6 million years ago and intensified from 875,000 years ago. Overprinted on this overall cooling trend, there have been dozens of more rapid climate swings between warm and cold. These changes in climate are due to natural alterations in the distribution of solar heat radiation around the Earth's surface. The dominant influence comes from the way the Earth's rotation and orbital characteristics vary in what are known as the Milankovitch cycles.

As a result of this periodic variation in solar heat distribution, there have been many cold and warm climate variations, termed 'stages'; there have been more than 100 stages in this ice age. During the coldest of these, thick ice sheets extended hundreds of kilometres onto lowlands, each remaining for only a few tens of thousands of years or less. The icy cold stages are known as glaciations, or 'glacials'. Warm stages between glaciations are known as 'interglacials' when the ice sheets retreated and occasionally disappeared entirely. During previous interglacial stages, the climate was sometimes even warmer than in the present day interglacial, with animals such as elephant and hippopotamus living in Essex. Each climate swing was usually very slow in terms of human existence but extremely rapid on a geological timescale. Over the last 875,000 years, cold periods have resulted in extensive ice sheets reaching into the area of the British Isles. Of these, only the Anglian glaciation 450,000 years ago resulted in an ice sheet that extended as far south as Essex.

The steppe mammoth, *Mammuthus trogontherii*, lived across Europe and Britain from about a million years ago. These were very large animals, reaching 4 m (13 ft) tall at the shoulders. The males had spiral tusks, with a recurved tip, that could grow as long as 5 m (16 ft). Our own Essex steppe mammoth features in the MIS 12 Anglian scene on the front cover of this book and at the head of this chapter. A population of smaller mammoths, derived from the steppe mammoth, evolved between 800,000 and 400,000 years ago in Siberia, becoming the woolly mammoth, *Mammuthus primigenius*. The steppe mammoth was eventually replaced around 200,000 years ago by the woolly mammoth, the last mammoth species to survive.

◄ Mammoth, elephant, rhino, horse and other fossils found in a post-Anglian interglacial layer at Aveley. *Trustees of the Natural History Museum London*

Climate change and ice age geology

Essex ice age geology provides an internationally important record of climate change. The surface geology of much of Essex consists of a veneer of sands, gravels and clays, together called 'superficial deposits' (shown as 'drift' on older geological maps). They were left behind by ice, winds, lakes and rivers throughout the current ice age. These layers, generally up to 40 m (130 ft) thick, but in places much thicker, provide evidence of our most recent geological past.

The superficial deposits of Essex are mostly the result of cold-climate action; only a relatively small amount was deposited during the many warm stages within the ice age. Nevertheless, some of these warm-climate deposits contain significant fossils and evidence of human occupation. Inevitably, information from earlier parts of the ice age is partly obliterated by subsequent glacial deposits and by continued river erosion, so usually the most recent deposits provide the clearest picture of events.

Ice was present during many cold stages through the ice age, going back millions of years. Stages when ice sheets and ice caps form are called 'glaciations' or 'glacials'. The first really large volumes of ice in Europe developed around 875,000 years ago. Yet it was only during the Anglian glaciation, through 450,000 to 430,000 years ago, that parts of Essex were reached by an ice sheet.

Great quantities of ocean water went into snowfall that piled up to form the ice sheets during each glaciation. This lowered the sea level each time, until much of the area of the present North Sea became low-lying plains. Because this happened during each glaciation event, animals have migrated back and forth multiple times across the area referred to as Doggerland. The ancestral River Thames flowed north-eastwards across Doggerland to join the Rhine and other major European rivers flowing north into the sea.

◄ The Thames and Rhine rivers at a time of low sea level, before the Anglian glaciation.

Ice age or glaciation?

Sometimes a glaciation, and in particular the last one, is referred to as an individual 'ice age'; however, this can lead to confusion when referring to the duration of the whole of the current ice age of the last 2.6 million years. To prevent such confusion, this book refers to these relatively short and icy later stages of the ice age simply as 'glacials', glaciations or glacial stages and not as 'ice ages'.

MIS: marine isotope stages and ice age climate changes

Microfossils in ocean mud provide us with a detailed record of ice age climate. Scientists have generated this record by analysing different types of atoms in the fossilised shells of tiny organisms, foraminifera, in ocean beds. The mud in the deep ocean contains a continuous record of climate change because the proportions of these different atoms in the fossil shells vary according to the temperature of the water the animals lived in. The ocean bed is drilled and cores of mud lifted to the surface for analysis. The climate variation results follow closely those of the Milankovitch cycles and are called marine isotope stages (MIS).

Different types of atoms – isotopes – have different atomic weights. The proportion of two particular oxygen isotopes varies with ocean temperature. For each level in a mud core sample, this proportion can be measured in the shells of the foraminifera. The result is a ratio of 'heavy' to 'light' oxygen, referred to as an 'isotope ratio'. The oxygen isotope ratio ($^{18}O/^{16}O$) in the shells reflects the ratio in the ocean water during the lifetime of the organism. 'Light' water, $H_2^{16}O$, evaporates more easily than 'heavy' water, $H_2^{18}O$. During colder periods, evaporated ocean water in the air is richer in 'light' water. This water condenses to fall as snow which is incorporated into the ice sheets. As a result, the remaining ocean water is relatively enriched in 'heavy' water and so its $^{18}O/^{16}O$ ratio increases. This ratio is reflected in the make-up of the shells of the foraminifera. The melting of the ice sheets at the start of the subsequent interglacial stage returns the 'light' water into the ocean, decreasing its $^{18}O/^{16}O$ ratio. This is again recorded in the shells. The record of oxygen isotope variation of countless foraminifera in deep sea cores is thus an indirect indicator – a proxy – for ice volume and climate change through many hundreds of thousands of years. Other research results, such as the analysis of atmospheric gas bubbles – including greenhouse gases – and dust within cores of ice from polar ice caps, support these findings through the past, particularly over the last 120,000 years. These cores also reveal many smaller changes in climate through hundreds or even tens of years, telling of a climate with rapid swings between warmer and colder conditions over very short periods of time.

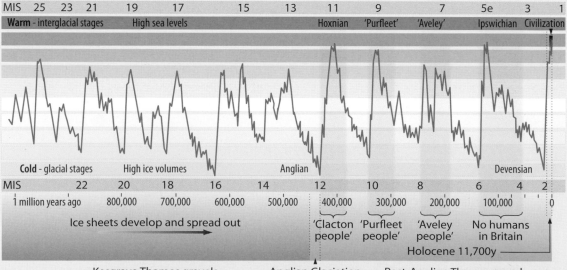

The major glacial and interglacial cycles represented by marine oxygen isotope variations are designated with marine isotope stage reference numbers (MIS #). Glacials have been given the even MIS numbers and interglacials have odd MIS numbers. Over 100 such stages, numbered from the present MIS 1 to the oldest MIS 100 and higher, have been identified over the last 2.6 million years. The graph produced from these data reveals the duration and magnitude of the variations. It also shows that the onset of warm periods is rapid, whereas the descents into glacial cold are more gradual. Particularly warm but short spells during a glacial stage are known as 'interstadials' and cold snaps during an interglacial are called 'stadials'. The latest cycle of glaciation uses MIS numbers 2 to 5, an exception to the general rule of MIS numbering due to the greater clarity of evidence for these more recent climate changes. Thus, during this last glaciation cycle, MIS 4 refers to a cold snap and MIS 3 to an intervening warm spell; there are also frequent archaeological references to MIS 5(a) to (e), these being warm (MIS 5a, 5c, 5e) and cold (MIS 5b, 5d) substages during the gradual cooling towards the deep freeze of this latest glacial stage, the Devensian, MIS 2.

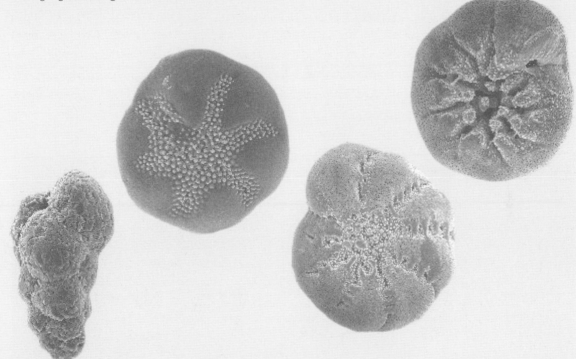

◄ Graph of climate stages in the last million years. The major hot and cold climate swings are numbered by marine isotope stage (MIS), making a handy reference for all the main variations in ice age climate. This is particularly useful when describing the geology of Essex, as many ice age stages are represented in this county's geology. We are living in 'MIS 1', or Stage 1, known as the Holocene interglacial. As well as the duration and magnitude of the variations, this graph shows that the onset of warm periods is rapid, whereas the descents into glacial cold are more gradual.

▲ SEM micrographs of ocean bed foraminiferans approximately 1 mm in size.
Wikimedia Commons

Ice age River Thames: the Pre-Anglian Kesgrave Sand and Gravel

The Thames river system was established as a major feature of the landscape tens of millions of years ago. The crust across Britain and into Essex was tilting upwards while the North Atlantic Ocean was opening out to the west. North-west to south-east rivers would have started to flow across southern England. During this time, another slow but powerful effect upon Essex came with a gradual collision of continents – that of Africa and Europe. As a result the Wealden area of Kent and Sussex and the southern part of Essex and the London area were compressed, gradually squeezing the rock layers. The landscape was altered, affecting the flow direction of the Thames. Beyond its channel through the Chalk of the Chiltern Hills at Goring Gap, the river flowed across northern Essex into East Anglia to the North Sea area.

From around 2 million years ago, riverbed gravels were being spread across the area of northern Essex and into Suffolk by the ancient Thames. Sand and gravel pits have exposed these layers, where they have been seen on top of the Chillesford Sands and older rocks. The Thames gravels in these areas are rich in flints that had been eroded in their billions from the extensive Chalk uplands to the west. These must have filled the river valleys and were gradually moved downstream by river floods that developed increasingly with intensifying cold climate stages.

The presence of quartzite pebbles, a very tough sandstone, in the earliest Thames gravels indicates that the river must have been eroding and transporting rocks from at least as far away as the west Midlands. As the uplands to the west of Essex continued to be eroded down, the river swept sands and gravels across Essex and Suffolk and the area of the North Sea in courses that changed with time. The overall effect upon the Earth's crust was to transfer the load of rocks from the west of Britain to the east and south-east, resulting in a change in balance that

▼ A large braided river showing flood gravel bars.

affects the lie of the land. As a result, the lands to the north-west of Essex have very gradually lifted. As sediments gather across the North Sea area, the crust has tilted downwards in the area of south-east Essex. This see-saw adjustment of the land of Essex continues to this day, albeit overlain by the effects on the crust of ice sheet advances and retreats on a shorter timescale, plus the rapid present-day effects of land use and sea-level rise.

The subtle effects of these processes on the landscapes of Essex have been recognised through research over many decades. Ancient Thames riverbed sands and gravels have been studied in Essex and beyond mainly in old gravel pits and big new gravel workings. The history of the river systems of southern England was revealed more clearly by extensive research in the 1970s. Descriptions of the precise pebble contents of each gravel, together with noting the various heights at which the layers occur, reveals a remarkable pattern of river development across the region.

As the ice age developed through the past couple of million years, especially during its glacial stages, thick snow covered the Essex landscape every year. As the snow melted each spring, enormous volumes of water eroded and transported rocks and pebbles along wide, shallow rivers. As each flood waned, the resulting gravel was left as 'bars'. The slowing water flowed in channels between the bars. Such rivers are now seen in high latitudes of Canada and Siberia where they are fed by seasonal meltwaters; often they are many kilometres wide.

▼ The Kesgrave Sand and Gravel was spread across Essex and Suffolk in several stages before the Anglian glaciation.

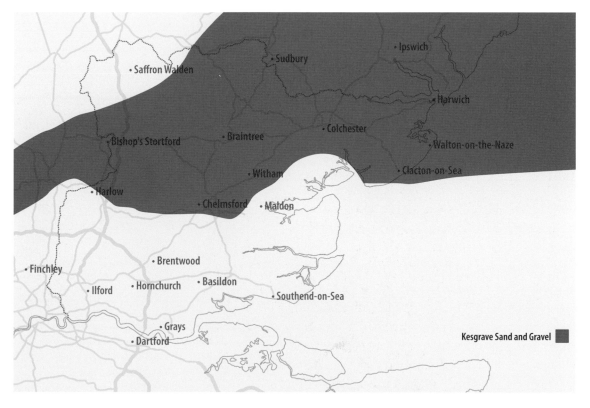

These early Thames riverbed layers are known as the Kesgrave Sand and Gravel. They were all laid down before the great Anglian glaciation, during which an ice sheet spread across Essex. Much of the Kesgrave Sand and Gravel is now hidden beneath a layer of glacial deposit left by that ice sheet. Today the ancient 'Kesgrave Thames' gravels are a valuable economic mineral resource and are worked as 'Essex White Ballast' in large quarries in the Harlow, Bishop's Stortford, Chelmsford, Braintree and Colchester areas. Kesgrave gravel pebbles and cobbles are composed largely of flint, but they also contain significant and fascinating pebbles, cobbles and even large boulders from even further west. The rarer far-travelled river materials are called 'exotics' to distinguish them from 'erratics', a term that refers specifically to materials transported by ice in glaciers and ice sheets.

Exotic rocks

Exotics provide clues to the headwaters of the rivers and the origins of their rock content. For example, volcanic rocks from the gravels provide an identical match with Ordovician ignimbrite and ash of north Wales. They are distinguished by their pale grey-greenish colour as well as their variety of volcanic textures. In particular, the gravels contain a large proportion, up to 20% in some areas, of quartzite pebbles and cobbles, a tough silica-cemented sandstone. This quartzite was originally derived from the mountains of Brittany in northern France; from there it was taken by desert wadi floods some 230 million years ago in the Triassic period, and transported northwards past Worcester towards Birmingham and Nottingham. Much later, the Thames floods eroded out many of these smoothly rounded, tough cobbles and pebbles and brought them across to Essex and beyond. These have a fine sugary texture of sand grains and they range in colour from pure white to liver-coloured, with varying amounts of iron staining. Many of them appear among the flints in the walls of churches throughout Essex and surrounding areas.

Exotic black-and-white rocks originally from the area of Cornwall were also eroded out of the Triassic pebble beds and brought to Essex. These beautiful rocks contain black tourmaline. Thames tributaries introduced white vein quartz, possibly from the parts of south Wales affected by mountain-building forces. Vein quartz pebbles are often pure white, showing a coarse crystalline structure and a degree of translucency. Pieces of silicified fossil wood are also found in these gravels, but their age and source have never been established.

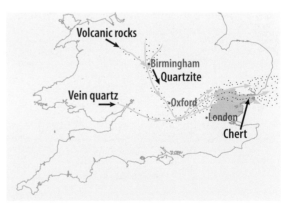

▲ Pebbles and rock fragments were dispersed among the billions of flints across Essex by the Thames and its tributaries. These 'indicator' materials provide clues to the origins of the rivers and the stages in their histories.

▲ Flints, quartzite and vein quartz in Thames gravel,
Bulls Lodge Quarry near Boreham. Exotic pebbles
provide clues to the origins of the gravel.
The overlying thick layer of glacial till
is seen in the background.

Quartzite (above) and vein quartz ►
from Thames river gravel at Boreham.
Quartzite looks sugary under a lens
and is opaque; the vein quartz
is slightly translucent and looks
more crystalline.

▲ Volcanic rock from north Wales and (right) the same type of rock found in the Thames gravel near Boreham, Essex.

Flint was derived from the eroded Chalk uplands to the west of Essex that were once much more extensive. The many well-rounded, distinctively 'chatter-marked', often black-coated flint pebbles were originally eroded in early Palaeogene times, from around 65 million years ago, laid down in beaches, and eventually re-eroded by the Thames system. Other, more angular flints were derived directly from the Chalk and are often white-coated.

All of these interesting pebbles and cobbles can be discovered on coasts, fields and footpaths and in many of our gardens around Essex. Peering into roadworks and holes in the ground can at last become an enlightening experience, as can the close examination of cobble walls. The many and various flints and a remarkable selection of exotic cobbles and pebbles are seen in hundreds of cobble walls in and around Essex, notably in the many Norman and medieval church walls built with material from adjacent gravel diggings and field clearing.

A southerly origin for fragments in river gravels is revealed by the presence of Lower Greensand chert in river gravels across Essex. This tough rock,

usually full of tiny sponge spicules, was brought into the Thames rivers by tributaries feeding in from the south, notably the River Medway from the region of Maidstone, where chert-rich Kentish Rag eroded and contributed this important clue to the origin of the river systems across the region. These fragments are from within the Thames catchment area, so are not classed as 'exotic'; nevertheless, they are a most valuable indicator for their region of origin and led to the discovery of the ancient River Medway flowing north-east across the area of east Essex.

Fossil wood found at Stanway. It is very hard, ► consisting of quartz that has replaced or 'petrified' the wood, cell-by-cell.

▲ Variety in flint – a typical Essex wall in Terling.

◄ Essex walls are delightfully instructive. At Broomfield Parish Church you can spot a variety of flints and exotic rocks including a rare Cornish tourmaline breccia pebble shown below.

Flint

Flint

Vein Quartz

Flint

Quartzite

Flint

Black and white 'Cornish' rocks in Essex

Occasionally, an exotic rock specimen is found in the Thames gravels that appears to be completely out of place. One such pebble, collected in 1987 from the Kesgrave gravels at a Chelmsford gravel quarry, was analysed and found to contain both tin and tourmaline; moreover, its rock texture matched exactly with many rocks encountered around the margins of the granite moors in Cornwall. Many such specimens have been spotted across Essex and East Anglia, as well as across the country to the west and the Midlands. In 1887 an account of quartz-tourmaline rock found near Felsted was followed by investigation in the 1890s showing that these rocks occurred in gravels from the south-west and west of England and across to East Anglia and Essex including Walton-on-the-Naze.

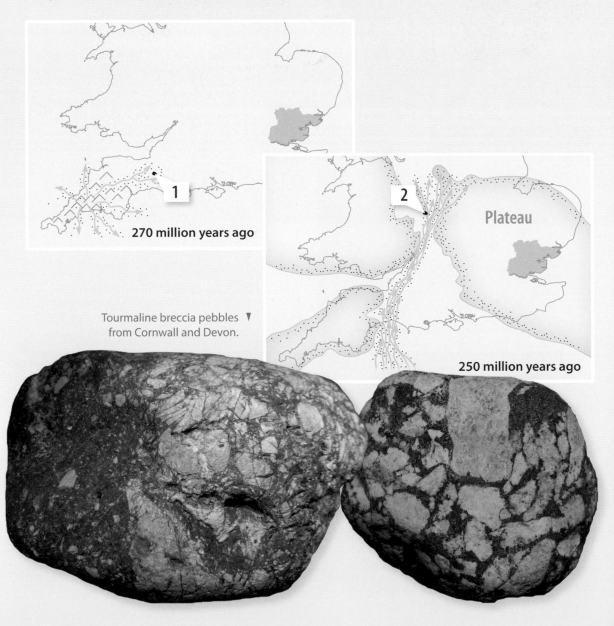

1 270 million years ago

2 Plateau

250 million years ago

Tourmaline breccia pebbles ▼ from Cornwall and Devon.

Tourmaline breccia pebbles ▲
found in Essex.

1 million years ago

◄ The sequence of pebble transport.

Only more recently has there been enough research into the origin and description of breccias and quartz-veined rocks in Cornwall, and also into the evolution of the Thames gravels in Essex, to reveal the likely story of these remarkable, very tough and beautiful pebbles. Many of these rocks are of breccia, broken-up rock fragments set in a finer groundmass. These match the 'explosion breccias' that formed beneath volcanoes or fumaroles around the cooling granite of south-west England 270 million years ago. Black tourmaline or 'schorl', an iron-boron silicate mineral, crystallised around the shattered rock fragments. The tough rock survived remarkably well as it was eroded from the Cornish mountains and transported across country through the long ages of geological time. The maps depict a Cornish pebble's 270-million-year journey, before it finally reached Essex from the west Midlands in the ancestral Thames.

River terraces

The Kesgrave Sand and Gravel layers have been spread by successive courses of the Thames as a set of large-scale terraces, like a 'staircase' across Suffolk and Essex. This feature is the result of two combined effects: the continued slow rise and tilt of the crust towards the south-east across the region, and the Earth's orbital variations affecting the climate.

At least eight terraces have been mapped within the Kesgrave Sand and Gravel, now almost entirely buried beneath glacial till. Correlation of these terraces with those of the Thames upstream of London enables some of the complex history of this great river to be told.

During the ice age, river flow has been at its maximum during each rapid warming at the end of a cold stage and, again, during the more gradual cooling into each new cold stage. At such times, seasonal snowmelt was at its greatest. There was little vegetation and the waters flowed unimpeded. The carrying power of the river was enormous and, at each flood, a broad new river course was formed, as depicted at the start of this chapter. Boulders, cobbles and sands were deposited in braided channels within the river. Between these episodes of high river flow, during each warm stage, the rivers were much smaller and, like the present-day River Thames, usually confined to a single sinuous channel which deposited relatively small amounts of mud and silt.

Each long period of glaciation was generally too cold for any appreciable river flow – the water was usually locked up in ice and permafrost. Yet the land continued to rise slowly throughout this time due to the persistent uplift of the crust beneath Essex. Consequently, the river in the next erosion period incised a new channel into the uplifted landscape. Because the land was also tipping to the south-east while it was rising, each post-glacial river torrent eroded into one edge of the previous flood plain gravel. As the cycle continued, a 'staircase' of terraces was created, the oldest being to the north-east and at the highest elevation.

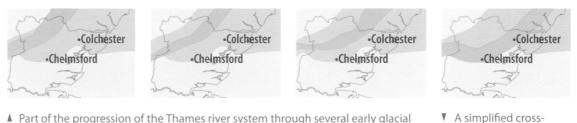

▲ Part of the progression of the Thames river system through several early glacial stages. The succession of gravels formed a 'staircase' of terraces – together termed the Kesgrave Sand and Gravel. The extent of these gravels across the area today is shown in pink although most of it is blanketed by later glacial deposits.

▼ A simplified cross-section of the Kesgrave terrace sequence across Suffolk and Essex, showing some locations that lie over each terrace.

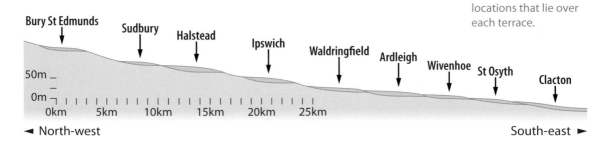

◄ North-west　　　　　　　　　　　　　　　　　　　　　　　　　South-east ►

The making of a gravel staircase

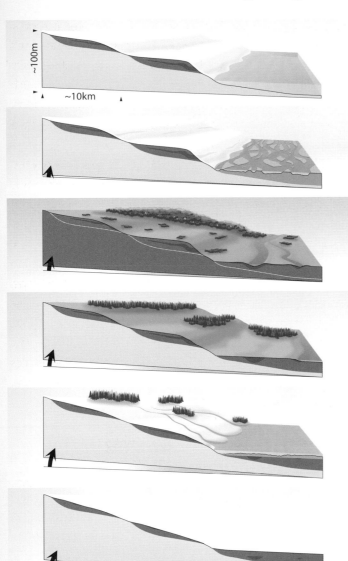

1. Big rivers eroding
As climate warms rapidly after glaciation, huge snowmelt floods cut a large new channel.

2. Big rivers deposit gravel
The river spreads sand and gravel along the channel bed.

3. Interglacial small rivers
During a warm stage, small meandering rivers deposit mud and silt in small channels.

4. Erosion
As climate begins to cool, the river cuts down into older sediment.

5. Big rivers deposit gravel
In the colder climate, snowmelt floods spread more sand and gravel.

6. Permafrost
Water is frozen and there is no river flow and little or no erosion while the land continues to rise and tilt.

The cycle repeats **Big rivers eroding**
As climate warms rapidly after glaciation, huge snowmelt floods cut a large new channel.

▲ Sequence showing how an Essex river terrace is formed over one complete ice age climate cycle of up to 100,000 years. The land is gradually rising and tilting from the north-east. Each terrace is a 'sandwich' of thin interglacial layers (black) within thicker cold-climate gravels. This cycle is repeated throughout the ice age and we are currently living in an interglacial episode, as shown here at (3).

◄ Small-scale modern
river terraces
in New Zealand.

▼ A valley incised into
a terrace near Mistley.
The terrace gravel lies
across the hill, top left;
a tributary to the Stour
has cut a valley into
the London Clay below.

Riverbed gravel

▲ This gravel layer is an ancient riverbed at Walton-on-the-Naze. Now 15 m (50 ft) above present-day sea level, it was laid down when the Thames-Medway river flowed across here to join the Rhine in Doggerland. The land has risen since then, hence the riverbed now lies along the top of the cliff.

The uplift beneath Essex has tipped the land up in the north-west through the past few million years, while the area of the southern North Sea sinks. The hinge-line in between lies offshore and parallel to the Essex coast. The rate of uplift has not been constant but its average maximum amount is in the order of a millimetre each 10 years, giving a general height of each terrace of around 10 metres per 100,000-year cycle. Each terrace is up to around 5 to 10 kilometres wide but they are largely hidden and preserved beneath the Anglian glacial till. This remarkable sequence of tilting and erosion that led to the formation of these river terraces has come to light only through several decades of painstaking geological fieldwork and analysis of gravels and their contents across the region.

Thames truncated

Analysing the proportions of the various pebble types in particular size ranges has enabled researchers to characterise the river gravels across Essex and East Anglia, helping them to interpret the history of the Thames river system. During the last 875,000 years, as the river cut down into the slowly rising landscape, the Cotswold Hills eventually formed a barrier preventing the drainage of the Thames from its former upper reaches in the west. The reduction in the proportion of volcanic rocks in later gravels shows that the headwaters from north Wales were cut off by the developing River Severn system, probably due to local glaciation in the Welsh mountains. Also, the

Thames headwaters were captured by the developing Bytham River, a river no longer in the landscape. As a consequence, the proportion of exotic pebbles in the terrace gravels changed through time and the younger terraces across mid-Essex differ slightly in their content of exotic pebbles from those of the older terraces further to the north-west. This was noticed after many stone-count analyses showed that the exotic pebble counts altered systematically. Thus, the terrace gravels across the area are divided into two distinctive sets, the older Sudbury Formation and the more recent Colchester Formation which contains a lower proportion of far-travelled pebbles.

▲ The Thames river system before and after cut-off.

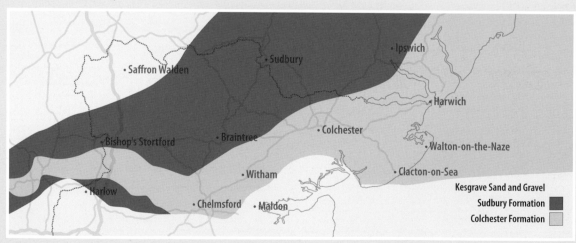

▲ The Kesgrave Sand and Gravel deposit, showing the earlier
Sudbury Formation and the later Colchester Formation.

Ancient soil layer

Across the top of each terrace is a distinct red soil layer which helps to distinguish the terrace in borehole records. During each warm stage the soil layer was formed in the top of the cold-stage gravel deposits that were left exposed. The iron minerals in the soil absorbed oxygen in the warm climate, giving the soil a red colour. Wind-blown silt and clay were also incorporated, forming a palaeosol (fossil soil) known as the Valley Farm Soil. The clay and iron oxide soil layers within the deeper parts of the soil profile are all that remain today.

This soil has been reworked during each successive warm stage. Also, during the cold stages, the soil was contorted by periglacial processes (described below), forming the Barham Soil. Thus, the palaeosol layer is complex and, as the soils on the older terraces went through more climatic cycles, their soils were even more affected. This means that the most complex soil profiles are on the oldest gravels, mostly in Suffolk. At Ardleigh and Wivenhoe the soil is less altered and at both localities it is associated with warm stage organic-rich deposits from small, muddy rivers. On the low terrace at Broomfield near Chelmsford there is evidence of a single phase of soil development.

▼ An ancient soil profile showing the warm-climate, red Valley Farm Soil overlain by later cold-climate Barham Soil that was affected by permafrost melting, becoming contorted. Cowlands Farm gravel pit, Stebbing (now under water). *British Geological Survey © UKRI 2021*

The River Medway in Essex

For much of the current ice age, the River Medway has been shared by Kent and Essex. The High-Level East Essex Gravels cap the hills at Hadleigh, Rayleigh and to the west of Bradwell-on-Sea. These gravels contain chert from the Lower Greensand that occurs in the north of the Weald of Kent. These distinctive Kentish pebbles enable the course of the River Medway to be traced northwards across east Essex to join the ancestral Thames around the area of Clacton. This chert is made of silica from sponge spicules and it erodes out as tough pebbles and lumps. When fresh it is bluish and slightly translucent, but it weathers to reddish or gingery-brown colours. With a hand lens it can be fun to spot the tiny spicules in these pebbles – they look like tiny white tubes or spikes.

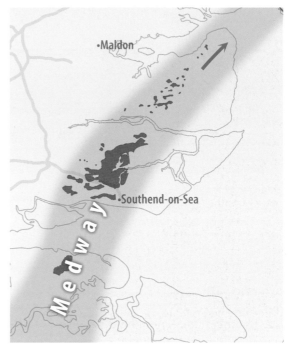

▲ The High-Level East Essex terrace gravels track successive courses of pre-Anglian Medway rivers that flowed within the blue area on the map.

▼ An unweathered block of Kentish Rag from the Lower Greensand of Kent, showing a band of dark bluish chert across the stone. Parish church wall, Hadleigh, Essex.

Medway Chert

Medway Chert

▲ Weathered ginger-brown chert fragments in gravel from the clifftop at Walton-on-the-Naze. These telltale rocks provide proof that river water from the Medway flowed across the area.

Greensand chert pebble. ▲

Sponge spicules

This rock contains distinctive spicules, the spike-like ► 'scaffolding' from sea-floor sponges.

1 mm

▲ Hills across east Essex follow the line of the pre-Anglian Medway riverbed. Beacon Hill, St Lawrence is an eroded remnant of the Medway course.

A close look into the ► bed of the Medway in Essex: Canewdon-St Lawrence Gravel at St Lawrence.
David Bridgland

Other Thames tributaries

Apart from the hills along the route of the ancient Medway, flint-rich gravels cap a number of hills across southern Essex, such as High Beach in Epping Forest, Brentwood and Warley, Billericay, Stock, Galleywood and Langdon Hills near Basildon. Dating these deposits, referred to as High-Level Pebble Gravel, has proved to be extremely difficult. There is still debate about their origin. The gravels were possibly laid down by rivers flowing north from the Wealden area across Essex and into the Thames along its course through north-west Essex and Suffolk. These southern sources of some of the High-Level Pebble Gravels are indicated by the presence of chert from the Weald, as with the Medway gravels.

Several metres of sand and gravel cap the hills around the Woodford area either side of the River Roding north-east of London. Lower in elevation than the High-Level Pebble Gravels of southern Essex, these are likely to be the bed of a river that flowed across the area at a later stage; nevertheless, this river must have flowed here before the Anglian ice diverted the Thames further to the south. The height of the gravels diminishes northwards and, together with the presence of chert from the Wealden area, this suggests that these too are the bed of a northward-draining tributary towards the ancient River Thames that flowed across north Essex and Suffolk. The Woodford Gravels have also been proved in a series of boreholes along the route of the North Circular Road past Woodford.

▼ Pre-Anglian courses of the Thames and Medway and other tributary rivers.

▲ Woodford Gravel at Knighton Lake in Epping Forest. *Diana Clements*

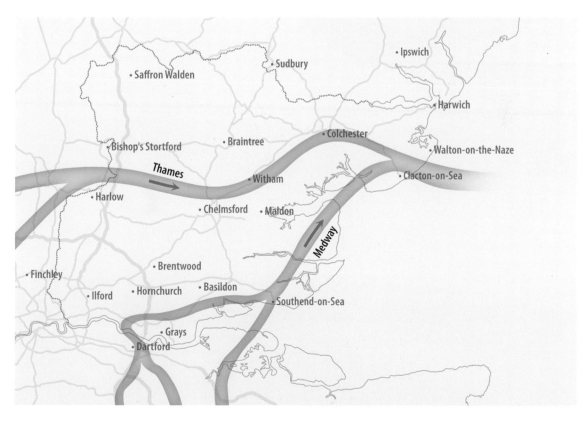

Riverbeds high and dry

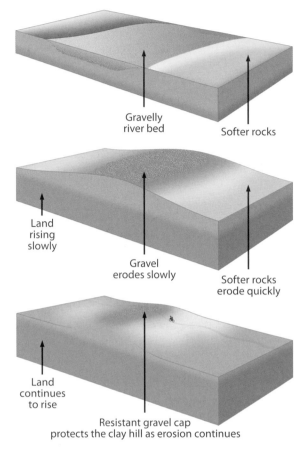

Gravelly
river bed

Softer rocks

Land
rising
slowly

Gravel
erodes slowly

Softer rocks
erode quickly

Land
continues
to rise

Resistant gravel cap
protects the clay hill as erosion continues

Some of these isolated patches of High-Level Pebble Gravel are found at elevations of over 100 m (330 ft) above present-day sea level; it is therefore difficult to imagine that each of these hilltops was once the floor of an ancient river valley. The hills owe their existence to the protective beds of river gravel that cap them. The gravel slowed down erosion during the gradual uplift of the region. During that time, the lands surrounding the valleys – without any such gravel protection – were reduced to the present lowland. The topography has thus been 'inverted' over time and there are now steep slopes down from the ancient gravel riverbeds. Because some of these hills are in the area between the Kesgrave gravel terraces in the north of Essex and the post-Anglian gravel terraces in the south and east, they stand in contrast to the unresistant clay and sands – devoid of gravel 'armour' – in the vale between these terraced areas. It is remarkable that so many hills in southern parts of Essex can make cycling such a challenge.

◄ Gradual uplift and erosion leave
an 'inverse-topography' of steep hills in south Essex.

◄ Langdon Hills, an example of inverted topography – an ancient riverbed lies along the crest of the hill 110 m (360 ft) above the fenland near Bulphan, seen in the foreground. Many pebbles of chert from the Weald of Kent are found across the hill top.

High-Level Pebble ► Gravel in an old quarry at Holden's Wood on the top of Warley Hill near Brentwood.
Gerald Lucy

Glaciers and ice sheets: the Anglian Glaciation across Essex

As temperatures fluctuated between glacial and interglacial over many hundreds of thousands of years, ice sheets periodically advanced and retreated. Within each glaciation, so much water was locked up in ice sheets that sea level fell greatly, sometimes as much as 120 m (400 ft) lower than today. The effect upon Essex geology and landscape has been profound.

During the most extensive spread of ice across Britain, known as the Anglian glaciation, an ice sheet up to 1 kilometre (over half a mile) thick reached as far south as Hornchurch around 450,000 to 430,000 years ago. The railway cutting on the Romford to Upminster line at Hornchurch is one of the most important ice age sites in Britain, providing evidence of this southernmost extent of the ice on mainland UK during the whole of the ice age so far. As the ice sheet moved across the land surface, it deposited 'glacial till', often referred to merely as till or, alternatively, 'boulder clay'. This superficial deposit covers a large area of Essex, having been left as a thick layer after the ice melted and retreated at the end of the Anglian glaciation.

The ice margin in mid-Essex, mapped from the remaining deposits of glacial till, was roughly along the line of the London to Colchester road (A12). Ice banked up against the Danbury-Tiptree ridge and the hills at Brentwood, Billericay and Galleywood. Tongues of ice penetrated further south to Hornchurch, Hanningfield and Maldon, probably along pre-existing valleys. The ice edge would have fluctuated from time to time over many hundreds of years; the ice was probably at its limit for around 5,000 years.

▼ The ice-margin across Essex mapped from what remains of the eroded glacial till deposits.

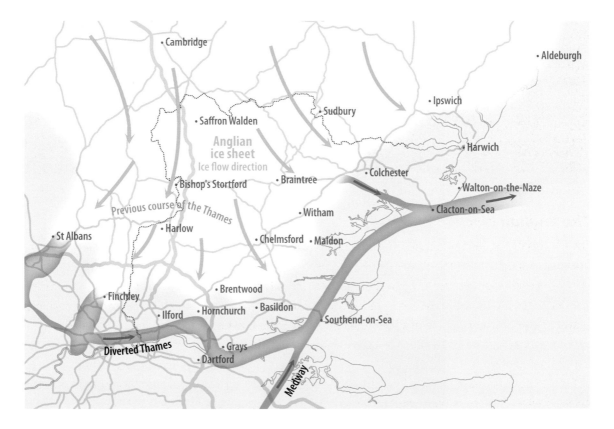

▲ A computer-generated view from 3 km (10,000 ft) above Billericay showing the edge of the ice sheet one summer 450,000 years ago. *Essex County Council*

The Anglian glacial till

The Anglian till has been studied in large pits around Chelmsford and Braintree and observed in numerous temporary excavations across the north of the county where it is seen to lie above the Kesgrave Sand and Gravel. The till is a bulk sample of all the rocks over which the ice passed, so its composition reflects that transported geology.

Essex till contains much chalk, so it is also known as 'chalky boulder clay'. In the north-west of the county

near Widdington it consists almost entirely of chalk. The lower part of the till layer is grey in places and is mostly derived from Jurassic clay as the ice crossed the Wash and Fens. The till layer contains a variety of rocks and fossils, such as ammonites and belemnites, although many of these have been broken or abraded by transport within the ice. There are also layers within the till where meltwaters inside the ice sheet deposited sand and gravel.

▲ Chalky till in Highwood SRC Quarry near Dunmow. People are standing on a thin red layer of glacial outwash covering the pale Kesgrave Gravel of the Thames riverbed.

◄ Large pieces of white chalk are seen embedded in Anglian till at Channels near Chelmsford.

▲ A scratched septarian boulder at Highwood Quarry near Dunmow.

As the ice moved across the land it ground up and carried along pieces of the rocks over which it passed, just as glaciers and ice sheets do today. Most of the particles were pressed as an unsorted rock mixture along the bed of the thick ice sheet. Sand and smaller particles stuck to other particles in the bed; larger fragments were lodged there and the ice dragged other fragments over these, grinding and scratching them. Glacially transported pebbles, boulders and rocks of all sizes often have scratches or striations over their surfaces, providing evidence that the rocks were scraped in the base of the ice sheet. The high-pressure 'sandpaper' effect of this process often smoothed and flattened the surface of the rocks, particularly chalk.

▲ Till plateau farmland near Broomfield showing the pale soil with chalk fragments.

Digging into ► the plateau. Glacial till overlying pale Thames gravel, Bull's Lodge Quarry, Boreham.

After the ice sheet melted and retreated, the mixture of rock material making up the glacial till was left in a thick layer across the land. Across Essex, the till usually consists of a fine mass of ground-up clay, limestone, sandstone, chalk and other rocks – with embedded small and large fragments up to boulder size, hence the term 'boulder clay'. The finest particles are not necessarily of clay mineral and the till does not necessarily contain boulders, so the term 'boulder clay' can be misleading. The plateau area of north and west Essex is extensively covered by Anglian till, mostly between 10 and 15 m (30 to 50 ft) thick, but in some places in the north of the county it can reach a thickness of over 40 m (130 ft). The till plateau itself provides excellent farmland, partly because of its high content of calcium from the chalk fragments in the soil and partly because there was a later dusting of wind-blown silt, the loess, from glacial-stage arctic winds, which also makes good farmland soil. The till plateau farmland soils were light enough for two-horse ploughing before the introduction of tractors, in contrast to the three-horse ploughing necessary for the heavy London Clay soils.

Where the till plateau has been cut through by later streams and rivers, pits were dug to extract the gravels below. Some of the valleys in the till plateau are quite deep, reaching the London Clay beneath the river gravels, and often these valleys shelter farms, homesteads and villages, notably where springs issue from the gravel layer.

The nature of the glacial till across Essex is variable and it has a number of different local names, like 'Springfield Till', 'Hanningfield Till' and 'Maldon Till'.

These could have been deposited during different advances of the edge of the Anglian ice sheet as the glacial climate continually fluctuated. The Anglian till sheet is termed 'Lowestoft Till' and, together with glacial outwash gravels and lake bed deposits of that time, the whole of the Anglian sequence of sediments is referred to as the Lowestoft Formation.

▼ Fossil ammonites in a large boulder of Jurassic limestone from the Anglian glacial till, Highwood Quarry near Dunmow.

Glacial erratics

Distinctive pieces of rock transported by the ice sheet and deposited as part of the till are called 'erratics', some from distant locations. This name is reserved for ice-transported blocks, whereas distinctive rocks that have been transported from some distance by rivers are sometimes referred to as 'exotics'. Essex till contains many glacial erratics. Before the lands of Essex were farmed, much of the till plateau surface would have been strewn with erratic stones and boulders as the till was weathered and eroded. Today, an interesting variety of erratics can still be discovered. Matching these rocks with known outcrops in Britain and even in Scandinavia often reveals the direction of ice movement. For example, the till across Essex contains dolerite from the Whin Sill in Northumberland. The large dolerite boulder now on display at the Bedfords Park visitor centre was found in a nearby gravel pit, having been brought from Northumberland by the ice sheet and then incorporated into the Thames river gravels. Some erratics are particularly large, such as the Jurassic septarian nodule, 2 m (6 ft) in diameter, now in the grounds of Saffron Walden Museum. However, the presence of some erratics can only be accounted for by a rather longer story of multi-stage transport by both ice and river.

The local direction of ice movement across different areas of Essex has been established by measuring the orientation of elongate pebbles within the till. Such pebbles tend to be aligned with their long axes parallel to the direction of ice flow. Pebble orientations are transverse to the flow direction where the ice edge was in compressive flow, such as when banked against a hill.

▼ Dolerite erratic at Bedfords Park
near Havering-atte-Bower.
Inset: the boulder's journey.

Whin Sill, Northumberland

Marks Warren Farm

Buried tunnel valleys

Across parts of Essex boreholes have revealed deep, steep-sided valleys cut far down into the chalk, sand and clay bedrock and now completely filled with glacial deposits. These buried 'tunnel valleys' were formed beneath the Anglian ice sheet as major drainage routes for meltwater. The gravel-laden water must have been under tremendous pressure to carve such deep channels under the ice and in some places the water was even forced uphill. In Essex, the largest example of a buried channel is the Cam-Stort Buried Channel which extends from Great Chesterford near the Cambridgeshire border south as far as Bishop's Stortford. It passes beneath the village of Newport where it is more than 100 m (330 ft) deep, almost half of this depth being below present sea level.

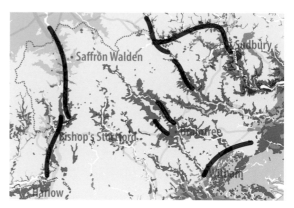

▲ Subglacial 'tunnel' valleys buried beneath the Essex countryside.

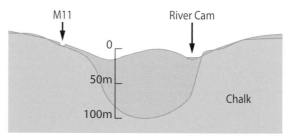

▲ A section across the Cam-Stort subglacial valley at Newport, near Saffron Walden.

▼ A mammoth tusk 1 metre across from beneath the gravel terrace across the Witham buried valley, Coleman's Farm.

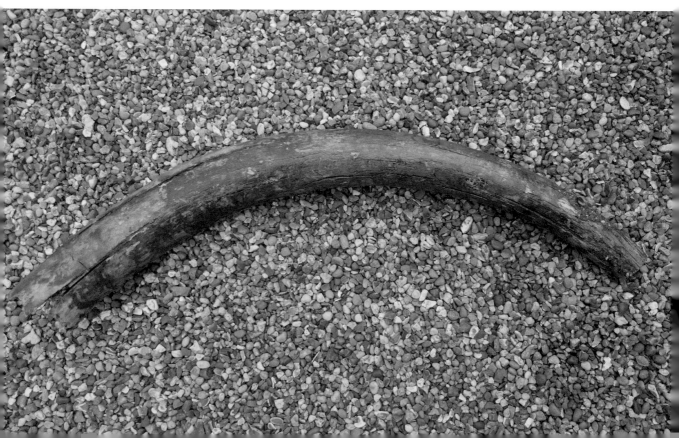

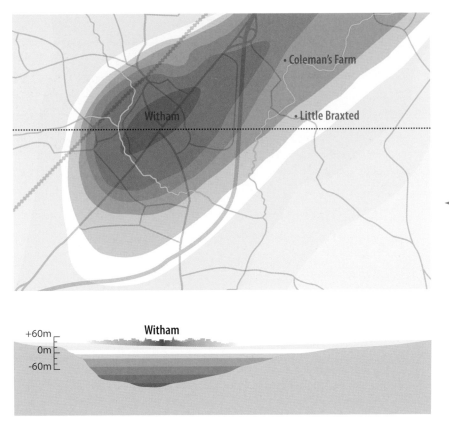

◄ A buried valley deep beneath Witham. This is an infilled channel that formed beneath the edge of the rapidly melting Anglian ice sheet 450,000 years ago. Overlying Blackwater river terrace gravels contain bones and teeth of large ice age mammals and flint artefacts made by humans.

The main road in Witham – the old Roman road, now Newland Street – lies 85 m (280 ft) above the bottom of a buried channel that formed beneath and along the edge of the Anglian ice sheet. A river under the ice scoured the rock below the ice edge at Witham and flowed to the north-east, uphill and under pressure. It carved through the Thames gravel and into the London Clay beneath the ice. The channel is at its greatest depth under Witham town, extending below present-day sea level. Here, the descent of huge quantities of water into crevasses in the melting ice margin rapidly deepened the channel beneath the ice. An upfolded layer of Palaeogene sands – beneath the London Clay – lies in the bottom of this channel, scoured by the river torrent.

The buried channel runs to the north-east from Witham past Kelvedon, beneath the present river Blackwater. A quarry within the channel at Colemans Farm, Little Braxted, reveals a deep filling of glacial till with small broken-up fragments of chalk and some sand layers. This was evidently washed into the subglacial channel by water from the melting ice sheet as the ice front pushed up against the slope towards the Tiptree ridge immediately to the south-east. Later river terraces of sand and gravel lie across the subglacial channel filling and it is these that are being quarried for much-needed aggregates. During quarrying, it was discovered that these terraces contain the remains of mammoths and other large ice age animals plus flint implements from humans living here in later stages of the ice age – a valuable window into diverse episodes of the ice age.

Around the ice edge

Around the edges of ice sheets throughout each glaciation, various deposits and cold-climate features were left in and upon the frozen landscape of Essex. These are referred to as 'periglacial' features, as they provide varied evidence of cold conditions affecting the area not directly overlain by ice, but mostly around the edge of the ice sheet and often extending 150 km (100 miles) or more beyond the ice front.

▼ The edge of an ice sheet. Braided rivers carry away glacial outwash. *Gerald Lucy*

Glacial outwash

Where meltwater flowed from the ice, an apron of sand and gravel was spread across the land by braided rivers in front of the ice edge.

In Essex, glacial outwash sometimes occurs beneath the Anglian till, having been deposited in front of the advancing ice sheet and then overridden by the ice. This thin layer of outwash deposits, known as the Barham Sand and Gravel, is seen as an orange-brown layer of gravels above the pale Kesgrave gravel in several quarries across Essex.

Stanway: an outwash delta?

In some places the outwash formed large fans of sand and gravel extending from the Anglian ice margin. Some of these deposits are preserved, as at Stanway near Colchester where they have been worked for sand and gravel for many years. As this sand is of glacial origin, and has been transported only a short distance by water from the ice edge, the grains are angular. It is thus a 'sharp' sand compared with the 'soft' sand with more rounded grains, resulting from longer transport of sands along the Kesgrave Thames riverbed. Detailed analysis of working faces and correlation of the gravel lenses within the sandier deposits has enabled a configuration of possible outwash delta channels to be mapped at Stanway. This has assisted extraction planning by the gravel company as well as adding to geological knowledge of the area.

At Holland-on-Sea, north of Clacton, works to restabilise the cliffs in 2018 revealed the Thames-Medway gravels above the London Clay. In the top of the cliff is a layer of gravel that contains more Medway chert and less of the typical Thames gravel content. The pebble content of this gravel reveals that it is a riverbed that was laid down at just the time the Anglian ice sheet reached its maximum extent north of Colchester. The gravel was laid down by the glacial outwash that was occasionally swept along what remained of the old Thames river valley south of Colchester. It is possible that the Stanway outwash delta also fed into this cut-off channel.

Meanwhile, the Thames itself was being diverted by the Anglian ice sheet to the west of Essex. The river course suffered a major 're-set', to flow instead across south Essex, joining the Medway close to Southend. The gravel in the top of Holland cliff is a rare marker of that enormous transformation in the history of the Thames and Essex. What is even more remarkable is that the rivers across Holland-on-Sea were, at that time, flowing towards a huge lake occupying the area of Doggerland just a short way to the east, where the southern North Sea is now.

▼ Sand and gravel layers in glacial outwash at Tarmac Quarry, Stanway.

▲ The Upper Holland Gravel exposed in the cliff in 2018. This layer is a mixture of Medway river gravel with glacial outwash along the former Thames channel.

◄ The outwash river from the Anglian ice edge continued along a short section of the old Thames channel, to join the Medway flowing from the south.

As the big glacial lake overflowed, huge quantities ► of meltwater from the Thames, Rhine and other rivers poured south through a new gap: the Strait of Dover was formed as an enormous waterfall cut through the chalk hills and fed a Channel River.

Huge flood severs Britain from Europe

The Anglian ice sheet extended across the lowlands from Essex to the Netherlands and Belgium. The southward flow of the ice sheet along Doggerland trapped the waters issuing from the Thames, Rhine and other rivers, preventing them from flowing along their usual routes to the north. The water had nowhere to go. At that time there was a continuous series of hills from the chalk Downs of Kent across to the Artois in France, forming a natural dam. Slowly, the rivers poured their summer meltwaters into the area to create a big 'pro-glacial lake' and the water level rose over hundreds of years until it eventually spilled over. The resulting massive torrents cut down into the soft chalk and drained away to the south over a series of vast waterfalls, forming a new gorge that became the Strait of Dover – the first severance of ice age Britain from the rest of the continent 450,000 years ago. The Essex Medway river, now joined by the diverted Thames, forged a new channel across the newly exposed lake bed, flowing south together with the Rhine and other rivers. A mighty combined river now flowed through the newly cut Strait and out along the English Channel towards the Atlantic Ocean. With a huge new river flowing to the south instead of to the north, it is likely that animal and human migration into Essex took place only across northern Doggerland after the melting of the Anglian ice and before the seas rose yet again, flooding the area to the east of Essex.

Since that great Anglian event, the sea has withdrawn and then flooded across the Doggerland area of the North Sea a further four times, while the climate alternated through its 100,000-year cycle of ice age heating and cooling. It is very likely that the Strait was widened out further during one or more of these later glaciations. Doggerland was each time a broad area of plains and lakes. The coastline of Essex is thus a very transient feature, varying in its present position, flooding and receding with changes in climate, until the next cold stage results in another departure of the seaside.

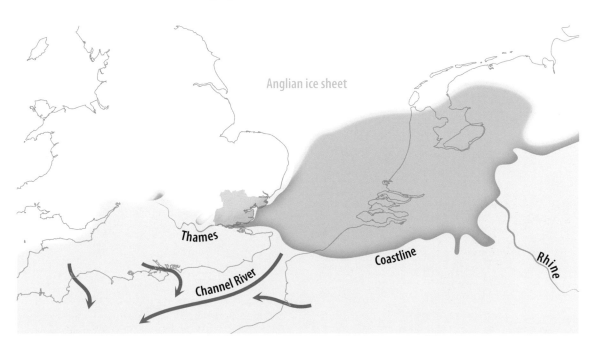

Outwash at Thorndon

At Thorndon Country Park, south of Brentwood, a 'Pebble Wall' section reveals an Anglian glacial outwash deposit in an old gravel pit. These sands and gravels contain large, rounded pebbles re-deposited from the nearby High-Level Gravel, together with a few erratics that can only have been derived from the ice 450,000 years ago, including Rhaxella chert from Yorkshire. The orientation of the pebbles in the lower layer indicates deposition in a rapid, braided-stream outwash flow from the north-west.

Close-up of tiny sponge remains in a Rhaxella chert ▶ erratic found in the Thorndon outwash gravel.

▼ Anglian glacial outwash gravel on view now in the lower part of the Ice Age Wall at Thorndon Country Park.

Ferricrete

The bright orange colour of the glacial outwash sands is derived from water-solution of iron-rich rock particles incorporated in the ice from further north. Iron is dissolved in the ground water which seeps downwards from the till into the underlying river sands and gravels. Iron-stained layers develop where the iron is re-deposited on and between the grains of sand and pebbles as brown to black minerals, largely of iron hydroxides and oxides – a kind of 'natural rust'. These rusty iron minerals cement the pebbles and sand together to form a natural concrete, hence the name 'ferricrete'. Blocks of this material, which hardens upon exposure to the air, are seen in the walls of many Essex churches. Some ferricrete is purely of iron-cemented sand, but much is of sand and pebbles, or broken pebbles, forming a ferricrete conglomerate. This is often mistakenly referred to as puddingstone, but has a different origin from the true silica-cemented puddingstone, a silcrete. The gravelly heathlands in Essex contain areas with 'iron pan' where iron from outwash, and also in some areas from the London Clay, has cemented gravels and provided a local source of ferricrete for wall building.

▲ Ferricrete in the process of formation in Kesgrave gravel, Highwood Quarry near Dunmow.
Ruth Siddall

Ferricrete ► conglomerate in the north wall at Ingatestone church.

Glacial lake deposits

The formation of lakes is a common feature of the margin of an ice sheet and there is much evidence for such lakes in Essex. The extensive ice sheet left numerous deep erosional and remnant ice features in the landscape. During late Anglian times and throughout the succeeding interglacial stage, sediment settled out from the lake waters forming distinct layers of fine sand, silt and clay on each lake bed. Such layers have been mapped on the modern land surface, seen in quarry workings for brickworks and logged in boreholes. The fauna, flora and climate records extracted from these accumulated lake sediments provide a valuable part of the great story of changing conditions in ice age Essex.

Where the extending ice sheet blocked an existing river valley, the river water ponded up against the ice front to form a 'pro-glacial lake'. There is evidence for this in the west of the county where the large Lake Hertford formed as the Anglian ice sheet advanced to the area between Epping and Harlow and blocked the course of the early river Thames across mid-Essex. As the ice continued to spread, lake water spilled over near St Albans and then at Finchley, so that the Thames re-established its course further south towards the line it follows today, as described later. This happened at

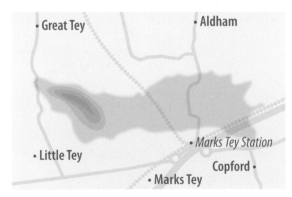

▲ The site of the ancient Marks Tey lake. Its clay bed provides a unique record of varying climates between 400,000 and 300,000 years ago.

the same time as the great lake was forming in the southern North Sea area, described above.

As conditions became warmer, meltwater from the ice sheet collected in hollows mainly in the till sheet, but also in valleys beyond the ice limit. Some of these lakes were probably too short-lived for sufficient flora or fauna to be preserved, so they are devoid of fossils that provide clues to the age of the lake sediment. The site of Lake Fyfield, near Ongar, has an estimated thickness of up to 20 m (65 ft) of lake deposits with some interbedded till showing that the lake was periodically buried by re-advances of the ice sheet. These deposits were used for brickmaking in the nineteenth century and earlier.

At Marks Tey there is a steep-sided lake basin situated in a hollow in the top of the Anglian till. Here, a lake existed for many tens of thousands of years. The microfossils within these deposits have been extensively studied, as they preserve a detailed record of evidence for the changing climate as well as for age dating. Pollen analysis was the main technique in the 1970s, but more recent work using ostracods – micro-crustaceans that evolved rapidly and are sensitive to climate change – is focused on both environmental changes and correlation with other warm stage deposits elsewhere in East Anglia. The succession of pollen and ostracods at Marks Tey reveals a transition after the late Anglian cold stage (MIS 12) around 400,000 years ago, through the succeeding Hoxnian interglacial stage (MIS 11) with a series of intervening warm and cold spells that occurred after the main interglacial, foreshadowing the next glaciation. Such a record has not been found elsewhere in the UK. From this detailed research it is clear that the uppermost part of the lake bed deposit records a very different climate from that recorded in the lower layers.

Limnocythere friabilis, found now in the Great Lakes of North America, suggesting a more continental climate; dominant in the uppermost part of the sequence, it records a very different climate from that of the lower beds.

Cytherissa lacustris, found in the UK today, as well as further north, is present throughout the sequence at Marks Tey.

0.5 mm

Fabaeformiscandona harmsworthi, currently found only in the Arctic so its presence at Marks Tey indicates periods of Arctic conditions in the earlier, lower beds.

▲ Ostracod microfossils and their record of climate changes at Marks Tey Lake.
Anna March/QMUL

▲ Lake bed sediment layers at Marks Tey. These clays are still used in brickmaking by W H Collier Ltd.

In the Chelmsford area, remnants of lake deposits are found in and around a buried channel at Sandon and between Boreham and Springfield, showing that glacial lakes covered these areas, most likely during the time when the Anglian ice sheet was melting.

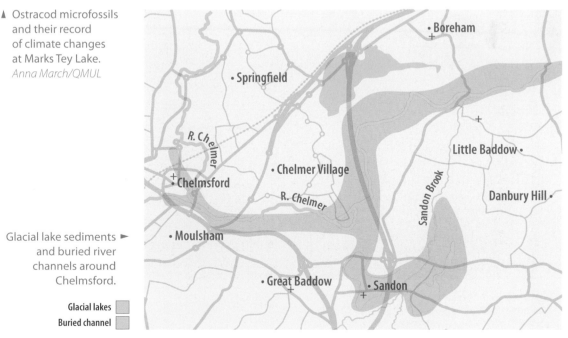

Glacial lake sediments ► and buried river channels around Chelmsford.

Glacial lakes ☐
Buried channel ☐

Frozen ground features

The freezing of groundwater accompanies the onset of each glaciation and this affects the nature of the land surface. The land becomes impermeable and the water from summer snowmelt flows across the surface rather than soaking into the ground, increasing the volume of the rivers. Only the top layer of soil thaws during each brief summer, the 'active layer', and this is subject to freeze–thaw action producing a variety of features characteristic of arctic soil.

Patterned ground

A characteristic feature of periglacial areas is frost-patterned ground. Today, these are sometimes revealed as crop marks in fields during hot, dry summers. Honeycomb patterns of polygons show where wedges of ice formed in the ground over years of freezing and thawing. As water freezes, it expands by 9% like ice cubes in the freezer, but when ice is frozen to below -15° or is subjected to a sudden drop in temperature, it contracts. Ice in the ground expanded and contracted annually making the cracks bigger, leading to infilling by ice in most cases, but by sand in arid areas, eventually creating ice wedges or sand wedges which penetrated the ground, often to a few metres in depth.

▲ Crop marks in the 1976 drought reveal periglacial patterned ground near North Weald Bassett. *Reproduced with permission of Cambridge University Collection of Aerial Photography, copyright reserved*

The Anglian periglacial ► landscape near Chelmsford 450,000 years ago. River plains with sparse vegetation provided a supply of dust for cyclonic winds to spread across the landscape. Summer melt from the snow and distant ice sheet brought great floods along the wide Thames valley through mid-Essex, leaving braided channels with sand and gravel bars. Summer melting in the permafrost resulted in solifluction lobes down gentle slopes and ice wedges opened out to leave polygon patterns in the river gravels.

Peter David Scott

Solifluction
lobes

Dust
storm

Edge of Anglian
ice sheet

Patterned
ground

Braided
river

▲ A 'cast' of an ancient ice wedge in glacial outwash gravel at Stanway near Colchester.
British Geological Survey © UKRI 2021

▲ An ice wedge in the Arctic today.
Wikimedia Commons. Ice Wedge

Older ice wedge casts, from the early part of the Anglian glacial stage, are sometimes preserved in the Kesgrave Sands and Gravels beneath glacial till and revealed by quarrying. During the most recent glaciation, the Devensian, the ice sheet extended only as far south as Norfolk, but the extremely cold temperatures throughout Essex at this relatively recent time have left fresh marks and features in the landscape.

Ancient soils

The widespread development of an arctic soil is recognised on the terraces of the Kesgrave Thames gravels. This is called the Barham Soil. It displays a whole range of frozen ground features and frequently overprints the older, warm stage soil, the Valley Farm Soil. In most places it lies directly beneath glacial till, indicating that it was the last surface process before the land was overridden by the Anglian ice. There is also some evidence of arctic soil development during earlier cold stages. Thus, the Barham Soil, like the Valley Farm Soil, cannot be attributed to a single climatic event. Rather, they each represent surface processes that reflect particular parts of the fluctuating climate cycles.

Red Barham Soil between the overlying Anglian till ►
and underlying Kesgrave Sand and Gravel of the
Thames riverbed. Highwood Quarry, near Dunmow.

Contorted bedding

Layers of sediment in the permafrost may become highly distorted after the ice melts during a spell of warming, notably at the end of a glaciation. This contortion occurs in layers where there is a lot of ice: as it melts, the particles of sediments are left unsupported in water and the layers are turned to quicksand. Heavier beds of sand may sink down, usually in lobes, into water-saturated silts and clays which have little coherence, so the beds become contorted. Pebbles can have their long axes rotated towards vertical by this process, providing further evidence of the thawing processes.

▼ Contorted bedding in sand and gravel layers near the clifftop at Walton-on-the-Naze.

Solifluction, head and soil creep

When thawed at the surface, the top metre or so of the active layer, over-saturated with water, is capable of flowing down a slope. This normally slow process, known as solifluction, produced a reworked deposit called 'head', the nature of which depends on which sediments were being transported. In Essex head is commonly found both on hillsides and in valley bottoms and locally forms an orange-brown, sandy, pebbly soil known as 'hoggin'. In the section at Thorndon Country Park, the layer overlying the glacial outwash river deposits mainly consists of reworked local Claygate Bed silty clay with occasional pebbles and this is interpreted as a head deposit.

▼ Solifluction lobes today, in Kyrgyzstan.
Marli Miller/geologypics.com

In some areas puddingstones and sarsens were transported down-slope from their original positions on high ground by freeze–thaw action. Water under the block freezes, lifting the block as it expands. Then, as the ice thaws the block is lowered, but as it is on a slope, it moves its position slightly downhill. Repeated freeze–thaw action over hundreds or thousands of years moves the blocks considerable distances. Many large boulders have been taken downhill into valleys by such action, notably the large sarsens and pudding-stones in north-west Essex and also huge sarsens in southern Essex along the edge of the Purfleet chalk ridge near Grays; the latter are certainly not glacial erratics as the area was never covered by the ice sheet.

▼ Head deposits beneath the surface soil around Chelmsford. These are areas of soft sediment which moved down gentle slopes during summer thawing of the permafrost – solifluction.

▲ The Pebble Wall in Thorndon Country Park near Brentwood: **Bed 1** is glacial outwash sand and gravel; **Bed 2** is 'head', a later, transported solifluction layer of Claygate Beds mixed with some gravel; **Bed 3** is a further solifluction layer with some vertically oriented pebbles and 'pot-lids' of ice-cracked flint; **Bed 4** is the vital, modern soil created out of the interface between geology and biology.

Head ▢

Frost-shattered sediments

Solid chalk is affected by freeze–thaw action. Water enters cracks in the rock, freezes, expands and widens the crack. When the ice melts, water makes its way deeper into the cracks and the process is repeated until the rock splits entirely. Frost shattered chalk, called 'coombe rock', becomes water saturated and, if there is a slope, the material may move downhill as 'head', or it may remain in situ. This process also makes the chalk more permeable, so that the valleys created by surface water when the ground was frozen are left dry as the water percolates down to flow underground through the rock.

▲ 'Pot-lid' frost-split flint pebble from a permafrost layer in the 'Pebble Wall' at Thorndon Country Park near Brentwood. This provides evidence of deep permafrost conditions.

▼ Permafrost-shattered chalk in a road cutting near Saffron Walden.

Some flint pebbles, despite their toughness, show characteristic breaks caused by freeze–thaw action. Flint is micro-porous and permeable to water. Rapid deep-freezing of this water sets up stresses within the pebble and, when repeated, causes the pebble to break from within. The break is often curved into the pebble, leaving a convex surfaced fragment and there are no percussion marks on the outside surface. These internally fractured flints have an indented part called a 'pot' and an ejected part, or 'pot-lid'.

Freeze–thaw processes are still active in the landscape, although on a much smaller scale than during glacial times. On freezing, ice crystals grow in the ground, lifting the soil and the stones. When warmed by sunshine, the ice below a stone thaws later than in the surrounding soil so that fine material moves in under the stone and holds it up. Over the years, stones are left protruding at the surface – 'Every farmer knows that the stones grow in their fields!'

▼ A dry valley in the chalklands of north-west Essex.

Loess and brickearth

During a glaciation, the vast river floodplains across the area currently occupied by the North Sea would have been covered by sand and silt dropped by the waters from each waning river flood. Large rivers such as the Rhine brought great quantities of dust from the ice-eroded Alps. Across the frozen, bone-dry and unvegetated plains, the finer sediments were picked up by circumpolar cyclones in ferocious dust storms. Redeposited as the yellowish dust and silt layers known as loess, thick layers lie from Siberia to the Central European plains and across to southern England. This process continues today with dust accumulating on cars after a few days during a dry summer – imagine what would accumulate over hundreds and thousands of years of fierce, freezing cyclones.

The loess dust is spread across much of Essex, but it is mostly too thin to be depicted on geological maps. Nevertheless, this dust-storm deposit is of fundamental importance as it contributes to the soils of the extensive farmlands in the county, its high calcium content enhancing fertility and influencing the soil structure.

Much of the loess seen in Essex was spread around the ice sheet during the most recent glaciation, around 20,000 years ago. With its distinctive 'columnar' structure of vertical cracks, a layer of loess is seen at the top of the low cliff along the Naze beach north of Walton, at Wrabness on the River Stour. The sandy loess layers across the Tendring area are mapped as 'coverloam' by the British Geological Survey and Soil Survey. This makes fertile soil ideal for growing grain and seed crops. Large spreads of loess also underlie Eastwood and the Southend area with thicknesses of up to 10 m. A small cliff is still visible at the old Star Lane Brickworks, Great Wakering, which closed in 2005.

In a few quarry locations in Essex, loess is also seen at the base of the Anglian till, showing the effect of dust storms around the advancing ice sheet of that earlier glaciation. Loess deposits arouse considerable concerns for engineering geologists, as their uniform grain size and poor cementing make a highly

◄ Loess deposit at Star Lane, Great Wakering.
Gerald Lucy

collapsible fabric. This can lead to very unfortunate foundation conditions and building collapse. In many areas of Essex, particularly in the south of the county, loess has been subsequently reworked by streams or hill wash and is commonly mixed with river sediment and redistributed London Clay. Some of these varied, silty geological mixtures were laid down by rivers during the most recent glacial stage during times of snowmelt, some 17,000 years ago. The deposits are often called 'brickearth', as they were formerly much used for brickmaking. They were extensively worked in the nineteenth century around Ilford and Grays, where the pits are now filled in. Significant finds of large and small mammal fossils occurred during the working of these pits, which are detailed later.

◄ A dust-storm deposit of loess is draped across the whole of the 'Naze Hill' at Walton. Here, in the down-slope cliff to the north, the buff-coloured loess rests directly upon the grey London Clay.

The columnar structure ► of the soft loess bed is seen beneath the soil. Just above the junction of the loess with the clay beneath, there are many small flints that may have been washed downhill by occasional meltwaters.

The Post-Anglian: a major diversion

The southern areas of Essex provide the best post-Anglian ice age geological record in Europe, if not the world. A remarkably complete sequence, with evidence of varying assortments of mammals through each stage of the ice age, of plants, human occupation, terrace gravel sequences and their interglacial 'sandwich fillings', all come together to give an increasingly detailed picture of the last half-million years of changes and migrations through to modern times.

The diversion of the Thames

Geologists have known for some time that a catastrophic change affected the Thames during the Anglian glaciation. The ice sheet, extending south into Essex, blocked the Thames valley upstream and diverted the river. Over the past 400,000 years, the diverted Thames has cut a further set of terraces south of the previously deposited Kesgrave Thames terraces, leaving a 'gap' in between the Anglian and post-Anglian deposits – a 'clay vale'.

These post-Anglian river deposits of the Thames between central London and the estuary are of unique importance within north-west Europe for their extraordinarily complete sequence, encompassing the last four climate cycles. In addition, a number of sites in north-east Essex contribute evidence for the evolution of the Thames-Medway river system. In particular, sites at Ardleigh, Little Oakley, Wivenhoe, St Osyth, Clacton and East Mersea have given us a picture of the changing geography of the area through the Anglian ice advance and into the following Hoxnian interglacial and subsequent stages.

Studies of the pebble content of gravels in a quarry at St Osyth and in the cliffs at Holland-on-Sea have revealed when, and just how suddenly, the Thames ceased to flow through central Essex. The Holland-on-Sea cliffs north-west of Clacton were uncovered in 2019 when the cliff was re-profiled to prevent landslips. The opportunity was taken to survey, sample and age date the gravels and results confirmed that the

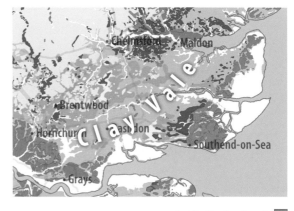

Post-Anglian river deposits ▮
Anglian: glacial till ▯
London Clay bedrock ▮

▲ The diversion of the Thames left a 'gap' in the superficial layers across southern Essex which is revealed as a 'clay vale' from Romford to Maldon.

lower gravels that rest on the London Clay are from a pre-diversion Thames at its confluence with the Medway. However, the upper gravels, with a greater proportion of material from the Medway than westerly derived Thames material, provide evidence of only a remnant river flowing from the ice sheet along a short length of the old Thames course and into the Medway, shown in the map of the ice sheet.

The clays and silts discovered in what is now the Vale of St Albans provide evidence of a lake, where the Thames was blocked by the advancing ice. The ice continued to advance until the water spilled into successive river valleys to take a new southerly route

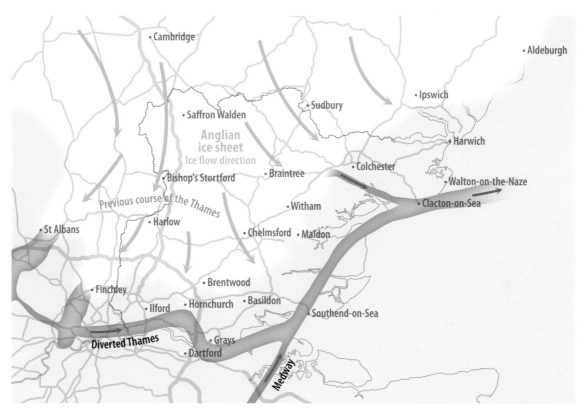

▲ The last pre-Anglian Thames through mid-Essex and the newly diverted river course.

▼ Essex and Doggerland showing the diverted Thames/Rhine rivers flowing south to the Channel.

to the sea. Painstaking detective work has enabled the changing courses of the river to be established with some accuracy by comparing the current strength and direction, and the distribution of rock types found in the various deposits.

During excavations for a railway cutting at Hornchurch in 1892, the oldest – and highest – terrace gravels of the diverted Thames were discovered to be lying on top of a remnant of glacial till, known here as the Hornchurch Till, which was deposited at the southernmost extremity of the Anglian ice sheet. Hornchurch is therefore an internationally important site, demonstrating that the present Thames valley downstream along southern Essex was formed *after* the Anglian glaciation.

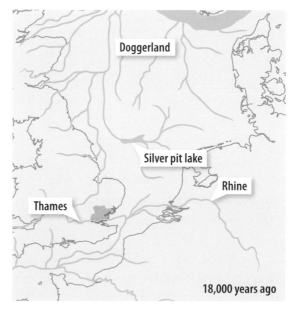

When the ice-diverted Thames reached the area of Southend it then flowed north as a combined Thames-Medway river along the old Medway valley towards the Clacton area. Evidence for this combined river lies in the Low-Level East Essex Gravel which contains not only characteristic rock fragments from Kent, but also a significant proportion of rocks from the west, brought in by the diverted Thames. At this time, when sea level was again very low and the plains of Doggerland stretched from Britain to the continent, the river flowed out across the area of Clacton to become a tributary of the Rhine. However, the huge, combined rivers now flowed to the south, through the newly cut Strait of Dover and then flowed west to the Atlantic along the line of the Channel.

Examining the pebble types ►
from the cliff to search for clues.

Riverbed sand and gravel at the exposed ▼
Holland cliff north of Clacton.

The Hornchurch story

The railway cutting adjacent to St Andrews Park in Hornchurch is one of the most important ice age sites in Britain. It is the type site of the Hornchurch Till, a deposit of glacial till laid down by a lobe of ice from the Anglian ice sheet along the Ingrebourne valley about 450,000 years ago. Hornchurch marks the maximum southerly extent of the ice sheet across mainland Britain during the whole of the ice age.

Here the till lies beneath the gravel of the 'post-diversion' Thames – the Orsett Heath Gravel of the Boyn Hill Terrace – and it is the only place where gravel of the *post*-Anglian, diverted Thames comes into contact with the Anglian till. The order of the strata – Thames river gravel on top of glacial till – in this one cutting is therefore the principal basis for considering that the entire modern Thames terrace system is younger than the main glaciation of eastern England. In other words, it proves that the Thames was diverted to its present course after the arrival of the Anglian ice sheet.

This discovery was made during construction of the Romford to Upminster railway line in 1892. The route required a cutting through a ridge of gravel-capped land running north-eastwards from Hornchurch parish church. The section exposed in the cutting was up to 8 m (25 ft) deep and 600 m (1/3 mile) long, showing about 5 m (15 ft) thickness of glacial till apparently occupying a depression in the London Clay and overlain by sand and gravel. The geologist and then vice-president of the Essex Field Club, T.V. Holmes (1840–1923), is credited with the discovery and his published observations and descriptions of the cutting in the *Essex Naturalist* and geological journals have proved invaluable. Holmes led a Geologists' Association field trip to the excavations in March 1892 and his field trip report states that many Jurassic fossils were collected from the till including a vertebra of a plesiosaur.

Re-excavation of a section in the cutting by geologists in 1983 revealed over 3 m (10 ft) thickness of till and confirmed that it is a typical 'chalky boulder clay', containing much chalk as well as abundant Jurassic rocks and fossils brought here from the area of the Wash and Fens by the ice sheet. The new excavation also confirmed that this site is of considerable significance for the correlation of the internationally important Thames terrace sequence with the glacial stratigraphy of East Anglia. For this reason, it is a geological SSSI, a primary geoconservation site. A further re-excavation of the section took place in May 2010 in advance of a television programme on the Ice Age in Britain.

▲ Anglian glacial till (pale) and the overlying Thames river gravel in Hornchurch SSSI. *Peter Allen*

Terraces of south and east Essex

The post-Anglian Thames and Medway terraces and sediments across southern and eastern Essex provide a remarkable record of the post-Anglian ice age, documenting the continuing 100,000-year climate cycle. These more recent terraces bordering the current Thames are free of thick overlying sediments, unlike much of the older and well-concealed Kesgrave Thames terraces further north. Consequently, the spreads of terrace gravels in south and east Essex, although eroded into patches, form several areas of flat and barren acid heathland such as Chadwell Heath, Becontree Heath, Squirrels Heath and Orsett Heath. The 'Flats' of East London, Leyton Flats and Wanstead Flats, are also examples of river gravel surfaces. The terraces of the post-Anglian Thames have provided London and Essex with expanses of land suitable for plenty of housing, industrial development and airfields over the past century, as well as the raw material for much of that building activity.

The terraces have been eaten into by subsequent river erosion and also much more recently by extensive quarrying. The dating and correlation of these terrace deposits has proved difficult because of their discontinuous and fragmentary outcrops. The gravels themselves are difficult to date, but careful observations have shown that the heights of the terraces reveal a sequence across the region. The warm stage sediments within the terrace gravels contain fossil evidence such as mammal teeth, shells and pollen that, with modern methods, have aided age dating. The staircase of terraces forms a subtle feature across the south Essex landscape. Looking south towards the modern Thames from the high ground of Havering-atte-Bower, steps can be seen in the landscape that help in visualising the whole staircase of post-Anglian river terraces.

▼ Terraces in the Essex landscape, looking south from Bedfords Park, Havering-atte-Bower. The names of the terraces help with correlation along the Thames Valley to the coast.

Top of the North Downs in Kent ►
River Thames ◄
Kempton Park ►
Taplow ►
Lynch Hill ►
Boyn Hill ►
Edge of Bedfords Park ►

Bedfords Park View

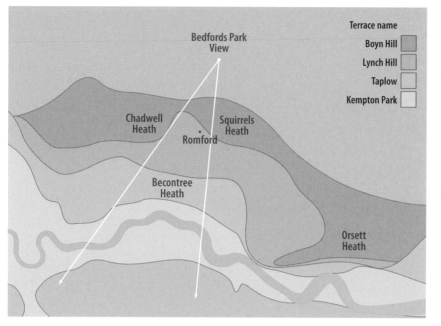

◄ The post-Anglian Thames terraces of south Essex. The view from Bedfords Park is shown in the photograph opposite.

▼ A cross-section of the post-Anglian Thames terrace 'staircase' in south Essex, showing their relationship with time and climate. *After David Bridgland /Quaternary Research Association*

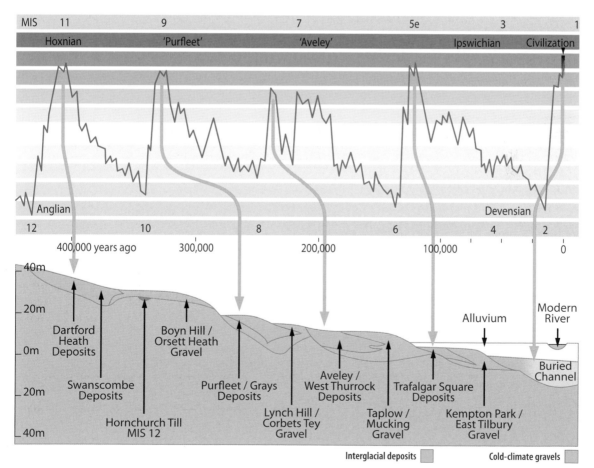

On geological maps and in field descriptions, various different names are used to describe the same post-Anglian terraces in Essex and the gravel deposits that form them. This can be most confusing. The table shows the names of the terraces, which are the flat stretches forming the landscapes of south Essex; it also shows the names of the gravels that make up those terraces. Each gravel is called after the place it was first recognised or described within the Thames valley. Warm stage deposits sandwiched within each terrace are also shown.

▼ A key to the post-Anglian terrace and gravel names and interglacials for south Essex.

Landform: Terrace	Geology: Gravel	Interglacial Stage	Glacial Stage	MIS	Approximate date (years ago)	Evidence in Essex
Kempton Park Terrace	E. Tilbury Marshes Gravel Formation		Devensian	2	12,000–100,000	Flint tools in south Essex Periglacial features
		Ipswichian		3–5e	120,000	Mammal bones at Trafalgar Square, East Mersea (Humans absent)
			(unnamed)	6	150,000	
Taplow Terrace	Mucking Gravel Formation	'Aveley'		7	200,000	Flint tools in south Essex Fossils at Aveley, Ilford, West Thurrock
			(unnamed)	8	250,000	
Lynch Hill Terrace	Corbets Tey Gravel Formation	'Purfleet'		9	300,000	Flint tools & fossils at Purfleet, LittleThurrock, Grays, Cudmore Grove
			(unnamed)	10	350,000	
Boyn Hill Terrace	Orsett Heath Gravel Formation	Hoxnian		11	400,000	Flint tools at Clacton Marks Tey lake deposits
			Anglian	12	450,000	Glacial till across north-west and mid-Essex

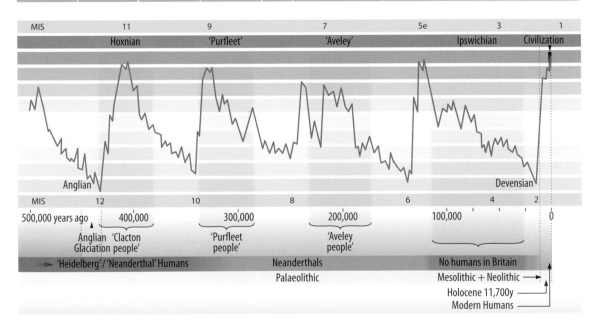

The gravel terraces of east Essex

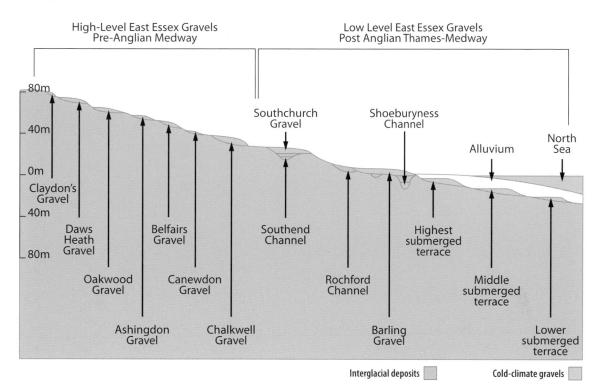

High-Level East Essex Gravels
Pre-Anglian Medway

Low Level East Essex Gravels
Post Anglian Thames-Medway

80m
40m
0m
40m
80m

Southchurch
Gravel

Shoeburyness
Channel

Alluvium

North
Sea

Claydon's
Gravel

Daws
Heath
Gravel

Belfairs
Gravel

Southend
Channel

Highest
submerged
terrace

Oakwood
Gravel

Canewdon
Gravel

Rochford
Channel

Middle
submerged
terrace

Ashingdon
Gravel

Chalkwell
Gravel

Barling
Gravel

Lower
submerged
terrace

Interglacial deposits ▢   Cold-climate gravels ▢

In the south-east of Essex, the High-Level East Essex gravel terraces, described earlier, reveal the existence of the Medway River across Essex up until the Anglian glacial stage. A little further to the east, the line of the Low-Level East Essex terraces contains evidence of both the Medway and the diverted Thames in their pebble content; they prove that the Thames joined the Medway from the time of the Anglian glaciation. These lower gravels also contain interglacial deposits in channels now buried beneath the surface at Southend, Rochford and Shoeburyness. The lowest terraces were submerged as the sea level rose at the end of the last glacial stage, the Devensian.

▲ Terraces – pre- and post-Anglian – and buried channels in east Essex. *After David Bridgland*

Thames and Medway low-level terraces and gravel ►
outcrops in east Essex, including the more recent
offshore gravel terrace deposits.

◄ Time, climate and human occupation
since the Anglian glaciation.

Evidence for temperate climates: interglacial deposits

The deposits of sediment laid down during the warmer stages of the ice age are both scarcer and of much smaller volume than the higher-energy deposits laid down during the much colder glacial stages. These warm stage deposits are mostly all of fine-grained sediments, some of them rich in organic material. Some consist of silty loam which originated in cold climates from wind-blown loess mixed with other sediments, later reworked by water as the next warm stage developed. These silt-rich layers are known as brickearth and 'loam'. Where they have survived erosion, they tend to exist in discontinuous patches. However, within each terrace sequence they contain significant flora and notable remains of fauna, including those of large mammals as well as evidence of human occupation.

Beneath the glacial till of north and west Essex, evidence of interglacial climates is seen across the surfaces of the gravel terraces in the layers of Valley Farm Soil. Within these pre-Anglian terraces, sediments, pollen and fossils have provided further clues to a warm climate, for example at Wivenhoe, Walton and Little Oakley. Yet within the post-Anglian terraces of the south and east, the interglacial record is relatively rich. Here, numerous small pits and quarries were dug in the past to extract their soft brickearth and clays for brick and tile making. These pits were originally dug by hand, which resulted in the discovery, and occasionally the preservation, of many bones and teeth of a large

▼ Some key sites where the interglacial 'sandwich fillings' of sediments have yielded a wealth of archaeological and geological evidence. These are found sandwiched within the colder period gravels of post-Anglian terraces across south Essex.

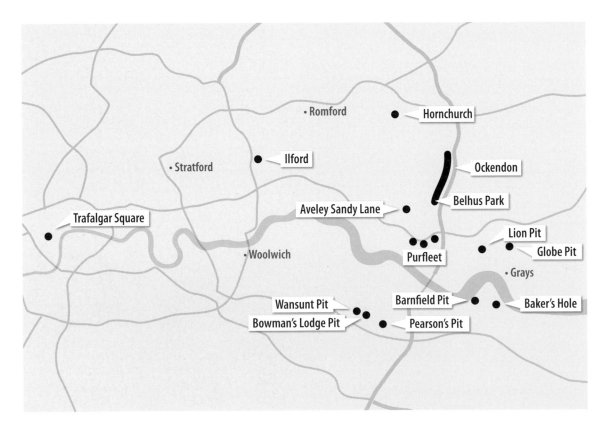

variety of mammals – from elephants to voles – plus many other life forms. The pits occasionally also revealed flint artefacts left by humans. Later, after large-scale machinery was introduced, fewer finds were discovered or preserved. However, a few spectacular discoveries have been made during and after modern quarrying and construction operations and during road and motorway construction.

Within south Essex, the Boyn Hill Terrace (MIS 12-11-10) is largely devoid of interglacial sediments. However, south of the present-day Thames, in Kent, the Boyn Hill Terrace includes the internationally famous site of the Swanscombe woman skull fragments, equivalent in age to the Clacton Channel site in Essex. The later, lower terraces of south Essex contain important interglacial deposits noted for their contents of mammal remains and worked flints.

Over 200 find spots in southern Essex make it a rich area for mammals and flints, and notably for Palaeolithic remains, over these interglacial stages. Remarkably, channel sediments belonging to four different interglacial stages occur in east Essex along the short stretch of coastline from Clacton to Mersea, providing evidence of climate and life in the Hoxnian (MIS 11), Purfleet (MIS 9), Ipswichian (MIS 5e) and the current Holocene interglacials. It is important that sites of previously undisturbed deposits that are to be excavated for mineral extraction or for construction projects are made accessible for geological as well as archaeological investigation.

Humans and large mammals

Human bones from earlier times than the present interglacial have not yet been found in Essex, but the flint tools that humans used – plus the flint debris from their manufacture – have been found in a number of the superficial deposits in and around Essex. Broken animal bones with cut marks from flint knives also provide dynamic evidence of human activity. Signs of occupation at various times during the ice age are rarely discovered and not evenly distributed across the county. Only layers of certain ages and certain types of sediment, and only those that happen to preserve the evidence, might yield any sign of humans. These layers are often obscured by buildings, transport infrastructure, farmland or industry. Added to this, the normal processes of erosion will have removed much evidence so that it is no longer available to us. Dredged gravels from the North Sea bed have been used to recharge beaches, for instance at Clacton; along these shorelines flint tools and worked fragments are often found as evidence of the occupation of the area of the North Sea – Doggerland – at the various times of low sea level. Human populations in the past may have been very small and isolated, even though the duration of occupation was across thousands of years.

Flint tools from the recharged beach at ▶
Holland-on-Sea, Clacton. The small flint,
5cm long, is Mesolithic (Middle Stone Age);
the larger two are Palaeolithic (Old Stone Age).
John Ratford collection

Mammal assemblage zones

Pollen analysis has played a significant role in describing and correlating the warm stages and compiling a sequence of ice age events. However, it has its limitations and much work now centres on the study of fossils of large and small mammals, invertebrates and microfossils. These all yield important information on climate and on time correlation. Within the sediments that were laid down during warm stages, geologists have discovered particular assortments – 'assemblages' – of mammal remains, bones and teeth, for each stage. Certain species appear only in beds of particular warm stages, never to return. For instance, at Clacton-on-Sea, the remains of the large fallow deer (*Dama dama clactoniana*) are unique to the MIS 11 interglacial. The teeth of rapidly evolving shrews and voles also provide useful clues. Thus, sets

Swanscombe MAZ

Cave bear
(*Ursus spelaeus*)

Large fallow deer
(*Dama dama clactoniana*)

Small mole
(*Talpa minor*)

Giant beaver
(*Trogontherium cuvieri*)

Rabbit
(*Oryctolagus cuniculus*)

European pine vole
(*Microtus subterraneus*)

Human artefacts
(Clactonian and Acheulean)

Purfleet MAZ

Brown bear
(*Ursus arctos*)

Water shrew
(*Neomys browni*)

Spotted hyaena
(*Crocuta crocuta*)

Modern fallow deer
(*Dama dama* ssp.)

Human artefacts
(Clactonian and Acheulean)

**Ponds Farm MAZ
& Sandy Lane MAZ**

Steppe mammoth
(*Mammuthus trogontherii*)

Horse
(*Equus ferus*)

Large Northern vole
(*Microtus oeconomus*)

Human artefacts
(*Levallois*)

Joint Mitnor Cave MAZ

Hippopotamus
(*Hippopotamus amphibius*)

Fallow deer
(*Dama dama dama*)

Absence of horse
(*Equus ferus*)

Absence of humans
(*Homo* sp.)

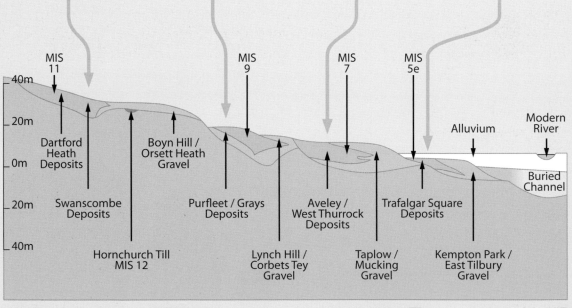

▲ Some of the fossil mammals that characterise the mammal assemblage zones.

Interglacial deposits　　Cold-climate gravels

of fossils characterise each warm stage and are called 'mammal assemblage zones' (MAZs). These provide a framework for dating and correlating deposits in and around the Thames valley. For instance, in conjunction with Thames terrace stratigraphy, they make it clear that the Purfleet, Grays, Belhus Park and Cudmore Grove river deposits were all deposited within the MIS 9 interglacial stage, around 300,000 years ago.

The use of MAZ evidence is becoming more sophisticated and it can now be used to identify mammal assemblages relating to climate peaks (interstadials) and lows (stadials) within an interglacial. For instance, through the 'Aveley' interglacial' (MIS 7), between 250,000 and 200,000 years ago, the earlier parts of the interglacial were characterised by woodland mammals and the later parts by grassland species.

◄ Mammal remains from the recharged beach at Holland-on-Sea, Clacton. Mammoth tooth, with rhino and horse. These could be of MAZ 7 age (200,000 years ago); however, research is continuing. *John Ratford collection*

Other aids to dating terraces and warm stages

Other fossils also provide valuable evidence for climate. Beetles are adapted to very particular ecological niches and fossilised beetle remains therefore provide very sensitive evidence of the climate and climate change at the times of deposition. Pollen found in sediments helps to build up a picture of the vegetation in the area over a particular period in time. A notable example of the use of this technique is the study of the laminated silts and clay at Marks Tey brickworks, the bed of a lake that existed right through the Hoxnian interglacial stage. Tiny fossil crustaceans known as ostracods have also become a most valuable means for investigating rapid climate change. The majority of these minute animals, around a millimetre across, live on or in sediments, and they are highly sensitive to their environment. With increasing knowledge of the temperature and oxygen demands and salt-tolerance of different species, the use of these environmentally sensitive microfossils in the study of ancient lake- and riverbed sediments of Essex has become increasingly important, as shown at Marks Tey.

'Clacton-on-Thames'

Along the coast at Clacton there is a set of channel deposits, a downstream continuation of the ancient Thames-Medway rivers. These channel sands and gravels form a record of the first deposits after the Thames had been diverted by the Anglian ice 450,000 years ago. This famous site, now obscured by sea defences, was discovered in the 1830s by Essex amateur geologist John Brown. Although Brown did not discover any worked flints there, the site has produced the bones of lion, rhinoceros and straight-tusked elephant. The flint tools found later provide the earliest undisputed evidence of human presence in Essex.

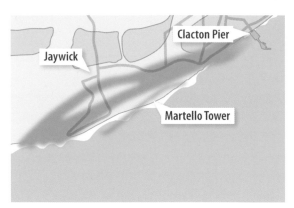

▲ The Clacton Thames/Medway channels beneath modern Clacton and its beach.

▼ The 420,000-year-old Clacton spear, discovered by Samuel Hazzledine Warren.
Natural History Museum/Wikimedia Commons

The channel deposits were extensively studied and described from the early Victorian period, when they were still accessible. In more recent times the few opportunities provided by redevelopment along the coast have enabled modern techniques to be used in continuing research. Flora, fauna and archaeological remains illustrate life in and around the river, spanning the end of the Anglian glacial stage and into the early Hoxnian interglacial, MIS 11. Higher in the sequence a later influx of marine species marks a change to estuarine conditions as sea level rose.

The deposits at Clacton provide a key link with the Thames deposits at Swanscombe and the East Anglian ice age gravels, making this one of the most important ice age sites in southern Britain and thus of international significance. The beds have yielded many Palaeolithic (Old Stone Age) artefacts, an assemblage of flint flakes and cores, but no recognisable tools such as hand axes. A working floor for making flint artefacts was found between Jaywick and Clacton. The perfect condition of the flint implements and the refitting of flakes back onto cores indicate that people were actually living very close to this spot, by the river, around 400,000 years ago. The Clacton flint artefacts represent a distinct Palaeolithic type,

recognised and named the 'Clactonian Industry' by Samuel Hazzledine Warren in 1911, a term which is now accepted and widely used. The tip of a wooden spear made of yew was found by Warren in the clays of the Upper Freshwater beds. It has been dated to 420,000 years and is the oldest wooden artefact found in Britain, now in the Natural History Museum, London. These Clactonians were not modern humans but were probably ancestors of the Neanderthals and possibly related to an earlier human species, *Homo heidelbergensis*. There can be no doubt, however, that humans also lived in and around Essex before the time of the Clacton people, before the Anglian glacial; maybe the evidence is waiting for discovery – perhaps within one of the earlier gravel terraces north of Clacton or in a pre-Anglian gravel terrace in the Southend area.

▼ Crudely worked flints from Hoxnian sediments at Clacton. These flints represent the earliest undisputed evidence of humans in Essex, approximately 400,000 years ago.
Drawings by John Wymer

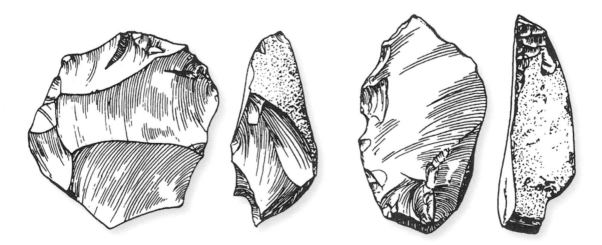

The sediments in the Clacton channel have provided a diverse fossil record, with molluscs from freshwater and estuarine environments and a vertebrate record ranging from bones or teeth of vole, water rat and beaver to boar, deer, horse, lion, rhinoceros and straight-tusked elephant, as well as the particular subspecies of fallow deer called *Dama dama clactoniana*. Pollen recovered from the freshwater sediments tells us that the surrounding countryside at this time consisted of oak, elm and alder forest, indicating a warm-temperate climate, which declined in warmth in the overlying estuarine beds as conifers and firs became dominant. Some of the molluscs are species that indicate the connection between the River Thames and the River Rhine – the 'Rhenish fauna'.

◄ Two views of a straight-tusked elephant tooth found at Clacton in the nineteenth century. The original owner of this tooth was living during the Hoxnian interglacial stage (approximately 400,000 years ago) on the banks of the Thames/Medway river. Despite its name, this elephant does have curved tusks but they are curved in one plane, unlike the steppe mammoth whose tusks are curved in three dimensions.
The Palaeontographical Society

◄ A skull of the fallow deer *Dama dama clactoniana*.
Natural History Museum /Wikimedia Commons

▲ The Cudmore Grove Gravel cliff, a cold-climate riverbed 280,000 years ago. *Trevor Johnson*

Mersea Island

At Cudmore Grove on Mersea Island, the cliff and beach reveal a section right through ancient river channels. Just under the beach a channel contains layers with cockle beds, overlain by freshwater sand with many fossils useful for age-dating, and a layer of peat with tree trunks. Fossils such as bones of beaver, bear and monkey have been found here, together with flint artefacts. Detailed research work has enabled the channel deposits to be dated to the MIS 9 interglacial stage, around 310,000 years ago. Pollen and ostracod microfossils helped to establish environmental factors such as climate and whether there was fresh or brackish water. The best evidence for the age of these interglacial beds comes from the teeth of small mammals such as voles and shrews, using

MAZ evidence. The channel, part of a very early River Colne or Blackwater which flowed into the estuary of the Medway-Thames, appears to be a small remnant of a large drainage network that established itself on the changing coastline of eastern Essex. It was a tidal river in a temperate continental interglacial climate with colder winters than today. Above the top of the channel, from beach level upwards, is a layer of gravel that is seen in much of the cliff. This gravel was laid down by a fast-flowing river during the succeeding cold stage (MIS 8) about 280,000 years ago.

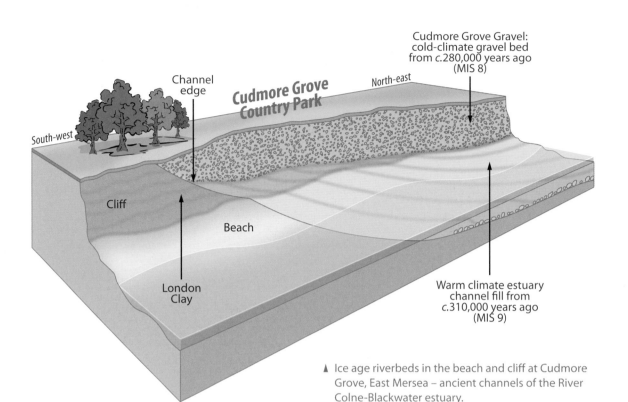

Cudmore Grove Gravel:
cold-climate gravel bed
from *c*.280,000 years ago
(MIS 8)

Channel
edge

North-east

Cudmore Grove
Country Park

South-west

Cliff

Beach

London
Clay

Warm climate estuary
channel fill from
c.310,000 years ago
(MIS 9)

▲ Ice age riverbeds in the beach and cliff at Cudmore Grove, East Mersea – ancient channels of the River Colne-Blackwater estuary.

Fossils of large mammals have been found at other localities on Mersea Island. At the East Mersea restaurant site, only 2 km to the south-west of Cudmore Grove, is another interglacial deposit, with silts and gravel channelled into the London Clay. It has yielded a freshwater mollusc fauna and fossils of ten vertebrate species including water vole and hippopotamus, a mammal assemblage combination unknown in Britain before or after that time. This assemblage compares closely with finds from several other Ipswichian (MIS 5e) sites from about 120,000 years ago. At another site on the beach, a distinctive fauna of the same age includes straight-tusked elephant, bison, giant deer and hippopotamus, referred to as the 'Hippopotamus Site'. It is difficult to imagine hippos wallowing in the Blackwater when collecting fossils on this often-wind-swept part of the Essex coast. Sea level was lower at that time; the Thames was flowing below its present

level and these sediments are from a previous course of the River Blackwater.

The interglacial deposit at Wrabness on the River Stour (possibly Ipswichian – MIS 5e, 120,000 years ago) yielded bones of elephant and mammoth at the beginning of the eighteenth century. They were described as 'diverse bones of an extraordinary bigness' and the writer concluded that they were probably bones of elephants brought over by Emperor Claudius for use in his wars with the Britons. Ipswichian fossils were also found at Walton-on-the-Naze, close to Walton town, and bones from here were the subject of the earliest recorded reference to fossils in Essex. Editions of Camden's *Britannia* (1610–1695) refer to the bones of 'giants' being found in the thirteenth and sixteenth centuries.

The Thames 400,000 years ago (MIS 12-11-10). ►
This river laid down the Orsett Heath Gravel.

Purfleet

At Chafford Gorges, where the old Grays Chalk Quarry cuts into the Purfleet ridge, the riverbed gravel of the earliest diverted post-Anglian Thames is seen along the pathways across the top of the ridge. This is Orsett Heath Gravel making up the Boyn Hill (MIS 11) terrace, formed by the Thames 400,000 years ago. Because the landscape of Essex has been slowly rising through time, this terrace is at a higher elevation than the Lynch Hill terrace seen in the Purfleet and Belhus Park excavations (described below). It may come as a surprise that you can be walking across an older Thames riverbed at the very top of this slope, whereas the gravel of the later Thames river is down the hill. It may help to refer to the earlier diagrams that show how the Essex terrace 'staircase' has formed over time.

Walking on the Thames riverbed of 400,000 years ► ago: Orsett Heath Gravel in the Boyn Hill Terrace, seen in a footpath across the top of the Purfleet ridge at Chafford Gorges.

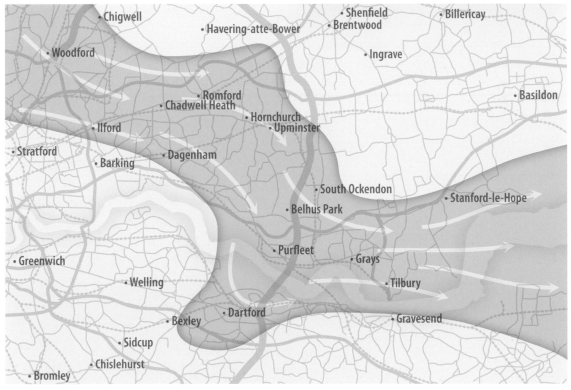

Several sites in the Purfleet area of south Essex together provide a complete sequence through the next stage of Thames deposits of around 300,000 years ago (MIS 9), linking with the Lynch Hill terrace deposits further west across London. Purfleet is one of a small number of sites in north-west Europe at which this interglacial period is represented. Chalk and gravel extraction in large quarries in the 1960s and 1970s revealed the cover of ice age sediments. In the 1990s, site investigations and archaeological excavations for the Channel Tunnel Rail Link (HS1) provided a well-funded opportunity to examine these deposits using the latest methods. A team of geologists, palaeontologists, archaeologists and geographers recorded and analysed material from all the available sites to record as full a picture as possible, placing the many Palaeolithic flint artefacts within a time sequence of interglacial deposits. Observations of the sediments at Greenlands Quarry some years earlier showed that they were deposited by a westward-flowing river. This led to an early conclusion that the river channel deposits here were from an early Mar Dyke tributary rather than the Thames. Further mapping over a wider area later showed that the channel here is, in fact, part of a large loop of the River Thames where it

had taken up part of the course of a previous river – maybe an earlier course of the River Darent where it flowed from the south and into the area around the Chalk of the Purfleet ridge.

As the land had risen further since the previous, 400,000-year-old Thames course, this new Thames incised its new course into the older riverbed. The edge of this newer Thames riverbed of 300,000 years ago is clearly seen here in a preserved section at Greenlands Quarry. Here, the riverbank is made of permafrost-shattered Chalk – coombe rock.

Within this riverbed – the edge of the Lynch Hill terrace – sands and shelly beds of the Corbets Tey Gravel contain large numbers of worked flints of Clactonian type near the base, followed by Acheulian and then prepared Levallois cores in higher beds. This sequence indicates that various waves of humans lived here on the banks of the Thames over a long span of time in human terms and, during this time, Neanderthal humans reached the area. The large amounts of worked flint reveal a picture of large-scale manufacture of hunting and food-preparation tools at an ideal site on a bend in the river, where plenty of fresh flint was available in the riverbank and along the riverbed. The steep bank made an ideal vantage point

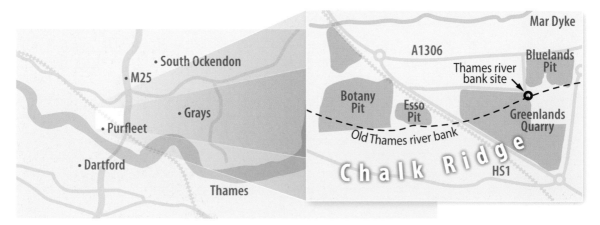

▲ Excavation sites along HS1 and surrounding quarries in the Purfleet ridge.

▲ The Thames riverbank of white, ice-fractured chalk, (coombe rock) with riverbed gravels preserved in Greenlands Quarry, Purfleet.

A braided river cutting into a ridge, similar ► to the Thames at Purfleet 300,000 years ago.

for game-spotting, perhaps for some thousands of years between 330,000 and 300,000 years ago. With a warm climate, high sea level and tidal waters reaching up the Thames to this location, European beavers, straight-tusked elephants and macaque monkeys lived in and around a knoll of grass and woodland. Fossil molluscs and the remains of mammals show that the Purfleet sand and gravel deposit at this site in fact represents a relatively short interval in terms of geological time within the whole of this MIS 9 interglacial stage.

▲ The M25 excavation into the 300,000-year-old Thames riverbed of the Lynch Hill Terrace, near Mar Dyke Interchange, 1979. *Essex Field Club*

▼ The course of the Thames around 300,000 years ago (MIS 10-9-8) showing the S-shaped 'Ockendon Loop'. This river laid down the Corbets Tey Gravel.

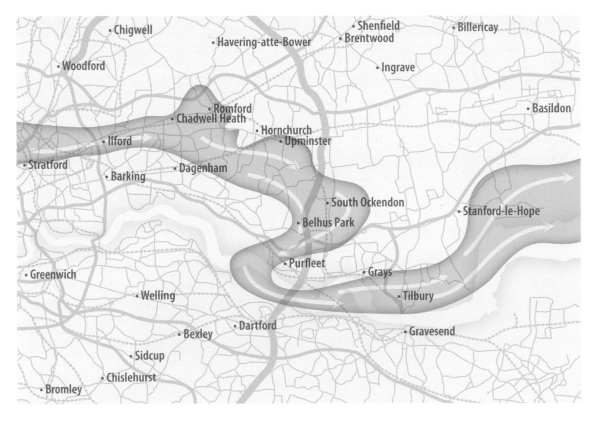

During the construction of the M25 motorway in 1979, excavations along the edge of Belhus Park revealed several hand-axes at the far bank of the channel from Purfleet, in the same ancient riverbed of the 300,000-year-old Lynch Hill terrace of the Thames, the Corbets Tey Gravel. One magnificent flint tool, 22 centimetres (9 inches) long and in very good condition, is now in the British Museum. Flint axes were not used for forest-clearing in those far-off times but were heavy-duty multi-use tools for cutting and food preparation. Perhaps this flint could have been a prize trophy – it is unblemished.

▲ A detail of the magnificent Belhus Park flint 'hand-axe' showing banded structure.

◀ A 22 cm (9 inch) flint 'hand-axe' butchering tool found in excavations for the M25 motorway at Belhus Park, Aveley. Photographed as found.
Essex Field Club

Further discoveries were made at Belhus Park when the motorway was widened in 2011. Analysis of the fossils and sediments provided more detailed information on age and climate. These were deposited at a time of high sea level during the 'Purfleet' inter-glacial stage, within a peak of warm climate referred to as MIS 9(e). Fresh-looking flint artefacts from this site were found close to their place of manufacture – evidence that 'Purfleet people' were busy just here around 330,000 years ago, working and hunting along the great S-shaped 'Ockendon Loop' of the ancient River Thames, on the shore opposite Purfleet. In these same layers were discovered water fern and water chestnut and – a very rare find in Britain – remains of a sabre-toothed cat, *Homotherium latidens*.

The fossil-rich brickearths of south Essex

The Lynch Hill Terrace (MIS 9) and Taplow Terrace (MIS 7) extend across a wide area of south Essex. Sandwiched within the gravels of each of these terraces are patches of interglacial sediments which have yielded countless bones of large mammals. Many famous sites have yielded mammoth and other spectacular fossils in the nineteenth century, such as those at Grays and Ilford, when brickearth was being worked for making bricks. Also associated with these terrace deposits are interglacial sediments channelled into the London Clay at Aveley, which yielded bones of a mammoth and a straight-tusked elephant in a clay pit in 1964 – and the first fossil found in Britain of a jungle cat, in a nearby road excavation in 1997.

Ilford's mammoth graveyard

In the 1850s, workers came across the bones of large mammals in and below the brickearth in Ilford's brick pits. The number of finds increased and they eventually came to the attention of Stratford amateur geologist Sir Antonio Brady. Mary Curtis, the wife of the owner of one of the brick pits, would send Brady a letter when the quarry workers uncovered any bones. The pits subsequently produced an enormous number of specimens, all excavated under Brady's supervision, but the greatest moment came in 1863 when the skull of the 'Ilford mammoth' was unearthed. It had tusks nearly 3 m (10 ft) long.

It is still the largest complete mammoth skull to have been found in Britain. Associated with it were the bones of a woolly rhinoceros.

Brady, a senior civil servant, devoted considerable time and money to ensuring that all the finds were preserved, and his collection was finally donated to the Natural History Museum in London. The collection contained the bones of more than 100 mammoths and at least 77 rhinoceroses. There were also bones of straight-tusked elephant, lion, brown bear and the giant deer *Megaloceros*, the span of whose antlers was a remarkable 3 m (10 ft). A catalogue of the collection was published by Brady in 1874 and a copy is preserved in the Essex Record Office in Chelmsford. The catalogue records the excavation of the Ilford mammoth in great detail and describes how difficult

▲ The skull of the Ilford mammoth, a late form of the steppe mammoth *Mammuthus trogontherii*. *Natural History Museum, London.*

the task was: 'You must imagine the skull resting half exposed in compact brickearth, requiring a spade or trowel to remove it, but the fossil itself as friable as decayed wood or tinder, the ivory of the tusks being equally soft and shattered.'

There is a remarkable abundance of fossil remains probably because they were carried by river currents and redeposited in the quiet waters of a meander. The Uphall Pit, on the west side of what is now Ilford Lane, was the most famous locality, but there were others to the north and south of Ilford High Road. During their working life the pits received many visits from naturalist societies, one of which was reported in the *Transactions of the Essex Field Club* of 1880 under the title 'A day's elephant hunting in Essex'. All the pits have now been filled in and the sites developed. Fossil bones do, however, still occasionally come to light at Ilford: in 1984 bones of mammoth, ox and rhinoceros were discovered during construction of Winston Way, the Ilford southern relief road. Mammoth and other fossils from the Uphall Pit are dated at around 220,000 years old (MIS 7).

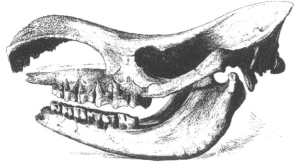

▲ The skull of a woolly rhinoceros from the Ilford brick pits. An illustration from the catalogue of the Brady collection, published in 1874.
Courtesy of the Essex Record Office

From the same 'Aveley' MIS 7 interglacial, at Chafford Hundred near Grays, there is an intriguing glimpse into human activity. In the old tramway cutting south of Lion Pit and Chafford Gorges, there is a section through what was originally a chalk river cliff, against which later deposits of sand and gravel have been banked up. Flint flakes have been recovered from gravels at the base of the cliff here. Some of these flakes could be fitted back together again to restore part of the original flint that was being worked by a human toolmaker. This refitting means that the flint flakes are still where the tools were being made.

Excavation at the Lion ▶ Pit tramway cutting in 1984. Neanderthal humans were seated at the base of the cliff at this very spot 200,000 years ago, making flint hunting tools.
David Bridgland

Two elephants and a jungle cat

In 1964 an amateur fossil collector, John Hesketh, discovered a number of large bones in a clay pit then being worked by the Tunnel Portland Cement Company Ltd on the north side of Sandy Lane, Aveley. Soon afterwards a team began a major excavation of the site and revealed two skeletons, one lying almost on top of the other. The upper one was of a mammoth preserved in a seam of peat, and the lower one was of a straight-tusked elephant preserved in silty clay. The mammoth is more recent than the straight-tusked elephant and they were found so close together because they were trapped at different times in soft marshy ground, apparently a silted-up channel. The sediment beneath the mammoth was found to be disturbed, probably by the animal struggling to escape. The skeletons were removed still partly embedded in a large block of sediment. They are now in the Natural History Museum in London. These animals lived during an interglacial stage around 220,000 years ago (MIS 7). Over 30 years after that discovery, in 1997, a nearby road cutting for the A13 exposed the same sediments and Aveley again found itself in the nationwide press. The reason this time was the first discovery in Britain of the bones of a jungle cat, *Felis chaus*, an animal that today lives in China, Central Asia and Egypt. Larger than a modern wild cat but smaller than a lynx, with a short tail and tufted ears, the jungle cat would have hunted small mammals, birds and frogs in the marshy area beside the river. With the bones of the cat were those of six species new to Aveley including a brown bear. The remains of many other mammals, including a version of today's lion, but 30% bigger, have also been discovered. Essex then bore a closer resemblance to the African savannah of today.

▼ The initial stages of the excavation of the Aveley elephant and mammoth in 1964. In the background is the clay excavating machinery belonging to the Tunnel Portland Cement Company, owner of the clay pit. *A.J. Sutcliffe*

The Devensian cold stage

Fossils from the Devensian glacial stage (MIS 2), the last glaciation, have been found at a number of sites across Essex. A gravel pit at Great Totham exposed a peat layer associated with the River Blackwater which contained bones of mammoth, reindeer and wolf. Devensian fossils have also been found at Essex ports, having been trawled up from the floor of the North Sea by fishing vessels; during the nineteenth century fishermen augmented their income by selling them to collectors. These fossils originate from a time when the sea level was much lower and the southern North Sea was again dry land – Doggerland – but today the bones lie in water depths of between 20 and 50 m (70–170 ft). Most of these fossils, which include fine mammoth tusks and teeth, have been taken to Dutch and other foreign ports to the benefit of collectors on the Continent. Also dating from this same stage are terrace deposits in the Lea valley which have yielded mammoth, woolly rhinoceros, reindeer and beds containing arctic plants.

When sea level reached its lowest level, during the coldest part of the Devensian glacial stage, about 25,000 years ago, the Thames cut a deep channel below the present Thames Estuary. The river extended out across the Doggerland plains that now lie below the southern North Sea. This channel still exists but is now filled with gravel and topped by recent alluvium in the bed of the present-day Thames and its estuary. This buried gravel channel is the downstream equivalent of the Kempton Park Terrace (MIS 2–6) west of London. Since that time, there was a period of warmth and then a sudden return to a glacial climate in Essex for a thousand years, during which time humans were probably still hunting in the area. Then came the rapid warming into the present interglacial stage, the Holocene – the subject of the next chapter.

The story of ice age Essex is not over. Further great swings in climate are to be expected, if not all so extremely rapid as the violent swing that we are currently experiencing – unprecedented during the current ice age. Some of the possibilities for the future may be discerned by looking into the past of Essex geology, through the ice age and before it. Our need for research and discovery continues. The role of the amateur geology enthusiast in searching, collecting and study in Essex is as important as ever in adding to this story.

10

The Naze Hill cliffs and Crag Walk sea defences at Walton-on-the-Naze in 2021. The historic tower is underlain by a thin layer of dust from icy sandstorms that blew from the frozen wastes of Doggerland during the last glaciation. Cliff-top layers are exposed as landslips develop, eating into the hill. As spring tides scour around the end of the defences the waves bite into the unprotected cliff and further slips develop.
Russell Wheeler

Quaternary	▼	**Holocene**

20,000 years	*Palaeolithic*	15,000		•	10,000	*Mesolithic*		•	*Neolithic*	•	*Bronze*	•	*Iron*	•	*Roman*	Now
▲				▲ ▲	Interglacial		▲ ▲									
Deep cold		Climate warming	Cold snap			Sea covers Doggerland			Farming							Global warming

Looking into the Essex landscape

After the ice

During the relatively short geological time interval of the ice age, there have been massive and repeated variations in climate, the effects of which were related in the previous chapter. Since the coldest period of the most recent glaciation, modification of the landscape has continued. In just the past few thousand years, the Essex landscape has been dramatically altered by human actions. There is much to discover from clues in the rocks and the land, a story to tell in every corner of the county.

The most recent glacial stage of the ice age, the Devensian, started around 130,000 years ago. The climate became progressively colder towards the deepest cold by around 20,000 years ago, with short interstadial periods when the climate temporarily warmed. Essex was not covered by an ice sheet during any part of the Devensian glacial stage, but it was a land of tundra and permafrost during the coldest periods, devoid of vegetation and overwhelmed by dust storms.

▼ The final, deepest freeze within the last – Devensian – glaciation was 20,000 years ago. It was followed by spells of rapid warming. Then a sudden return to a glacial cold snap lasted for about 1,000 years. A final warming 11,700 years ago, followed by a huge rise in sea level, established the present interglacial stage – the Holocene.

The Neanderthal people had become extinct in Europe by around 30,000 years ago, although they were not in Britain at that time. Meanwhile, the first modern humans began exploring the hunting grounds across Doggerland and into Britain. Around the time of maximum cold, there is no evidence of human habitation in the whole of Britain. Then, when the climate began to warm from around 15,000 years ago, modern humans again migrated across the game-hunting plains and around the fishing lakes of Doggerland – and they must have ventured into Essex. Their flint tools, found in the caves at Cheddar for example, are the last of those classified as Palaeolithic (Old Stone Age) artefacts. Woolly mammoths were still living in Essex until around 12,000 years ago and their teeth and tusks from this time are occasionally found in coastal sediments.

Frozen cold suddenly returned for a period of just over a thousand years from 12,900 years ago, affecting the whole of north-west Europe and the British Isles. It happened when the melting of Arctic ice spread fresh water into the North Atlantic Ocean, stopping the warming effect of the Gulf Stream in its tracks. This cold period, which brought back tundra conditions, is known as the Loch Lomond Readvance or the Younger Dryas. The winter temperature across Doggerland and

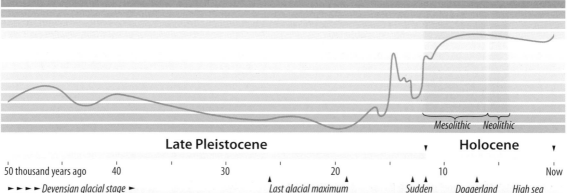

Mesolithic   Neolithic

Late Pleistocene   **Holocene**

50 thousand years ago   40   30   20   10   Now

► ► ► Devensian glacial stage ►   Last glacial maximum   Sudden   Doggerland   High sea
                                        + lowest sea level   cooling event   drowned   level

▲ The tusk of a woolly
mammoth in
sediments about
12,000 years old,
found in East Mersea
off Coopers Beach
in 2018.
*Mersea Museum
/James Pullen*

A woolly mammoth tooth 28 cm across from ▲
beneath the gravel terrace over a buried valley,
Coleman's Farm, Witham.

Essex averaged minus 17°C. Evidence from this time, in the sands and gravels along the Thames in Essex, is of sparse vegetation together with insects that now live in northern Scandinavia. This was a land of sedges, liverworts, mosses and low-growing trees – willow, juniper and birch. Summers brought fast-running floods, braided streams and rivers across a lonely area with few land animals – foxes, bears, wolves, and summer migrations of horses and reindeer. Some humans, probably in hunting groups, migrated with them, surviving the harsh times. Their long-blade flint tools have been found in sites to the west of London and in Kent.

Doggerland floods again

That last period of deep cold ended quite suddenly from 11,700 years ago, reaching a temperate climate within only about two centuries. The rains came, and with them came the greening of the land. The humans returned along the rivers and the Thames led them from Doggerland and along the wide valley between Essex and Kent and across where London is now. The 'edgelanders', the survivalists, followed the herds, always on the move.

Many different types of record, such as pollen and insects as well as ice cores from Greenland, all mark this warming event. It is taken as the end of the Pleistocene Epoch and the start of the current Holocene Epoch – the present interglacial. Great changes in the Essex landscape came with the migration of new plant and animal life into the area, while forests spread rapidly across much of the county.

Further bands of people migrated into the area from around 11,000 years ago. The rapid sea-level rise following the melting back of the Arctic ice sheet would have affected hunting, fishing, food gathering and migration each year. Between 8,500 and 8,000 years ago, the Doggerland plains were cut through by the sea for the last time (so far), and the coastline gradually receded until it came to resemble that of today. The climate has remained fairly constant since that time, with only minor colder spells from 5,500 to 5,000 years ago, again at around 2,500 years ago and through the 'Little Ice Age' of around 400 to 200 years ago.

The study of human remains and artefacts from prehistoric times, their associated animal and plant remains and the sediments in which they are found are the archaeologists' field of interest. A key aspect in the modern understanding of human history has been the relatively recent multidisciplinary approach to excavation and investigation, with various geological specialists, alongside archaeologists and geographers,

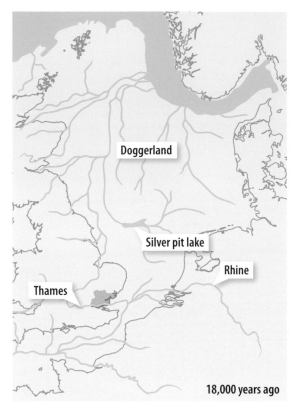

▲ Doggerland 18,000 years ago.

integrating their knowledge and experience to arrive at the maximum possible understanding of this dramatic story.

Much of the Thames floodplain is covered by alluvial deposits that overlie the sands and gravels laid down in the earlier, colder times of the Devensian stage. The Thames, the Blackwater and the Crouch-Roach river systems are estuaries that were drowned as sea levels rose and flooded inland. The saltings of the Essex coast and estuaries contain flint tools from the Mesolithic (Middle Stone Age) – from the end of the last glaciation almost 12,000 years ago to around 6,000 years ago. Mesolithic flint implements such as tranchet or heavy axes, known as 'Thames picks', have been found at Rettendon, Rawreth and Canewdon. Small flakes or microliths that may have been attached

▲ Doggerland hunting area 10,000 years ago. Throughout the seasonal migrations across this broad plain, hunting, fishing and foraging sustained human populations for many hundreds of years.

▲ Dogger Hills became an island – now it lies beneath the waves as Dogger Bank fishing ground beneath Dogger Sea Area, familiar to those who enjoy listening to the Shipping Forecast.

to wooden handles have been found at Mayland and Steeple along the Blackwater and Althorne on the Crouch as well as inland at Rettendon and Rawreth. Together with these, many other flints worked into chopping tools, scrapers and blades have been found, especially around the Essex coast.

Neolithic polished axes and arrowheads were made by an influx of people from 6,300 years ago who spread farming into Essex for the first time. Later societies brought bronze manufacture to the area by 4,200 years ago, succeeded 2,750 years ago by iron smelting. These periods are chronicled by the artefacts that were left and the interpretation of the geological context in which they are found. Rocks, sediments and soils were exploited for agriculture, building materials and pottery making.

Today's interglacial, the Holocene Epoch, is different from all that came before. This time, the human population didn't remain in sparse groups of hunters and communities gathering food and fishing. In this interglacial, the area of Essex and east London depicted in this book has become the settled home of more than 4 million humans. Geologically, this episode of human development and impact on the land is extreme, and unprecedented. We are living through a very sudden, if not catastrophic, ecological change and the lands and coasts of Essex reveal the extent and suddenness of these changes. Reading the landscape has become an important part of the recognition of what is happening – and preparation for what is to come.

Wormingford Mere: a remnant permafrost feature

Evidence for the latest melting of the permafrost is contained within Wormingford Mere, a deep, dark body of water close to the northern border of Essex. It is fed by springs and is now joined on the northern side to the River Stour through a small cut. It is surrounded by trees and fenced for use by a private angling group. This could be the site of a collapsed 'pingo', a melting permafrost feature now seen in Siberia where the permafrost is collapsing right now.

A pingo is a dome-shaped hill formed in a permafrost area when the pressure of freezing groundwater in one spot pushes up a layer of frozen ground. When this ice melts at the end of a glaciation, the drape of surface sediment across the mound collapses into the deep hole that was previously filled with solid ice. A deep circular lake is left, which may fill with fine sediment very gradually over the following few thousand years.

▲ Wormingford Mere may be the site of a pingo, a dome-shaped hill formed in permafrost, in the last glacial stage to affect Essex. Now a private fishing lake, it is just inside the border with Suffolk, between Colchester and Sudbury. *Google Earth*

▲ A newly collapsed pingo in Siberia today. *The Siberian Times*

A core of sediment was extracted from Wormingford Mere by Cambridge University Department of Botany in 1981. The results showed that in the centre of the mere the water was 6 m (20 ft) deep. Below this they found 9 m (30 ft) of soft dark mud, underlain by a further 5 m (17 ft) of dry, very compact mud. This lower mud was very difficult to penetrate, and eventually they could extract no further core samples from this depth even though they had not reached the bottom. Such a thickness of mud is unusual in English lakes. Analysis of the pollen and spore content of the mud 15 m (47 ft) below the lakebed showed that the mud at that level is about 9,000 years old and that the forest vegetation around the mere was very different from anything we see today. The landscape was dominated by hazel and birch with small amounts of elm. The team expected a further 4,000 years of record to be present in the bottom of the mere, extending back to the end of the last glaciation at least 13,000 years ago. Sadly, further research has not been possible. Such work throws valuable light upon the nature and rate of climate warming.

Fossil forests

During the deepest cold of the Devensian glacial stage, the coast was far away from Essex, with the sea level some 120 m (350 ft) lower than today. With the warming from 11,700 years ago, the cold, fast, braided flow of the River Thames gave way to a smaller, slower, meandering flow of muddy water along a single channel. Braided channels were cut off and filled with organic sediments and washed-in sands. Tundra gave way to extensive pine forest across London and south Essex through to 9,300 years ago. In the area of Silvertown, a filled channel contains pollen and sediment that shows these changes were followed by deciduous forests, then a decline of elm, and evidence of Bronze Age forest clearances locally from at least 3,000 years ago.

Sea level recovery was not constant. At times of higher sea level, mud was deposited along the Lower Thames shores, while during times of slightly lower sea level, vegetation, including trees, spread out across the mudflats. With the next high sea level, the vegetation would die off and become sealed beneath new layers of mud. Today these changes may be seen as a sequence of clay layers from the mudflats and peat beds from the vegetation. One of these peat beds, with tree trunks over 5,000 years old, can be seen by the sea wall close to the RSPB visitor centre at Purfleet. A further sea-level rise 2,500 years ago buried a peat layer in the Silvertown channel beneath more layers of estuary mud. Observations of such sediments reveal a story of natural changes superimposed by changes induced by human interactions with the landscape.

▼ Part of a submerged forest, between 5,000 and 6,000 years old exposed at low tide on the Thames foreshore at Purfleet. The forest was in existence in the Neolithic Period, at a time when sea level was much lower. *Bill George*

Standing stones

Before farming came to Essex, much of the land would have been littered with stones and boulders, mostly hidden within the wildwood. As forest was cleared and fields were tilled, an annual task for a farming community would have been to clear the farmland of stones and boulders, a practice that continued until relatively recent times. Tilling the soil was made easier if there were few stones to impede the plough. The stones would have been gathered to make lanes and trackways more passable and, eventually, they were much used for building 'cobble walls'.

The calcium-rich soils of the glacial till plateau of north Essex made good farmland – but they were also strewn with glacial erratics, the stones and boulders transported in the Anglian ice sheet 450,000 years ago. The clay soil shrinks and swells with wetting and drying and frost heave so that, every year, more

stones seem to 'grow' out of the soil. The larger sarsens and puddingstones, the legendary 'growing stones' or 'breeding stones', have become part of Essex lore. These stones were either regarded with suspicion and broken up in fear of them growing out of control, or they were so revered that they were removed to meeting points, cross-ways and boundaries, or to more social or sacred locations including stone groups or circles. In valleys and gravel pits in the till areas, the underlying Thames gravels also yield large boulders and these, too, have been removed and distributed as special stones and they still are to this day.

Most of these large stones are of sarsen, a sandstone, and puddingstone, a sandstone-flint pebble conglomerate. They are both silica-cemented sedimentary rock known as silcrete, rather like an extremely tough sort of natural concrete. Such tough, cemented sandstone

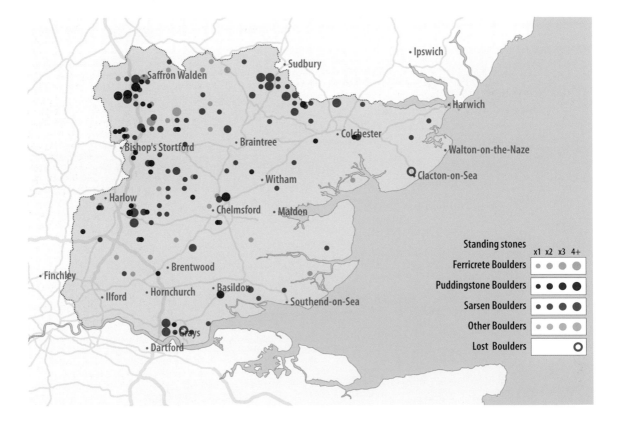

is known geologically as quartzite. In both south Essex and north-west Essex large silcrete boulders, many of them slab-like, have been concentrated by repeated episodes of freeze–thaw action – solifluction – during past glacial episodes, during which they have gradually slithered downslope from their place of origin. In south Essex, these are found along the Purfleet ridge and the Mar Dyke. The Anglian ice sheet did not reach this southern area, so these stones are not glacially transported erratics. Here also outcrops of the Upnor Formation containing sarsen and puddingstone have been directly quarried and some of the stones have been used in walls and churches, the largest being too massive to shift far.

Puddingstone and sarsen slabs have also accumulated in valleys in north-west Essex. These boulders are not glacial erratics either – the Anglian ice sheet went around the Chilterns – but were similarly moved by solifluction from sites where they originated. The highest concentration is on the flanks of the Chilterns, very close to the Essex border with Cambridgeshire. There are other gatherings of large stones in nearby valleys, for example in the village of Arkesden where the war memorial consists of a single large puddingstone. There are other puddingstone and sarsen slabs in the stream in the middle of the village which may have once paved a ford. Along the outcrop of the Lower London Tertiaries further to the east, the church at Alphamstone appears to have been built within a large stone circle of sarsens.

Across the till area of Essex, true glacial erratics of sarsen and puddingstone are occasionally seen as standing and marker stones. A notable example, once used as a boundary stone, is at a triangle junction along Dagnets Lane, Black Notley.

▲ Puddingstone war memorial, Arkesden churchyard.

▲ Sarsen erratic boundary stone inscribed to indicate a detached part of Notley parish. Incised 'Whit Notly' on the north side, 'Black Notly' on the south with the date '1679', this is a rare survival of a named and dated boundary stone in Essex; Grade II listed.

▲ Sarsen stones at the junction of Fryerning Lane in Ingatestone. © *Martyn Copcutt 2013*

Two notable sarsens guard the entrance to Fryerning Lane in Ingatestone. However, many large sarsens and puddingstones have been extracted during quarrying from gravels of the ancient, pre-Anglian Thames riverbed, from gravel pits and quarries across the whole of the county, with notable examples excavated at Stanway near Colchester and at Broomfield near Chelmsford.

A very few standing stones are of ferricrete, dark brown 'iron-pan' – iron-cemented pebbly sand. Less tough than silcrete, it tends to age-harden and has been much used for wall building across Essex. The Millennium Stone at Navestock is a giant boulder of ferricrete excavated from a nearby field. It is 2.1 m by 1.5 m (7 ft by 5 ft) in size and was placed by the road in 2000 to mark the Millennium. It is mistakenly inscribed as being a 'puddingstone'.

▲ The Millennium Stone at Navestock, a large ferricrete boulder. *Gerald Lucy*

◄ Puddingstone at Channels, along the Chelmer Valley at Broomfield, the site of a large gravel quarry. The plaque celebrates the high quality of land restoration.

Puddingstone querns

Dome-shaped quernstones for flour production were made from puddingstone for a short period early in the Roman occupation of Britain. These high-status devices, rarely more than 350 mm in diameter, were very difficult and time-consuming to shape, but the result was a very tough stone with a grinding surface that was slow to wear and which presented many sharp facets in the exposed sections of flint pebbles.

An iron drill bit using a rotating machine with a charge of crushed puddingstone would have been needed to make the central hole in the upper stone. The lower stones were usually more roughly finished than upper stones. Evidence for quarrying and production has been found in Hertfordshire, although finishing was likely to have taken place close to the areas where the stones were used. These remarkable quernstones were produced mainly between 50 and 100 CE, after which the flour production market was flooded with imports of Mayen lava from the Continent. A wide scatter of puddingstone querns has been discovered across eastern England, with a concentration across Hertfordshire and north Essex.

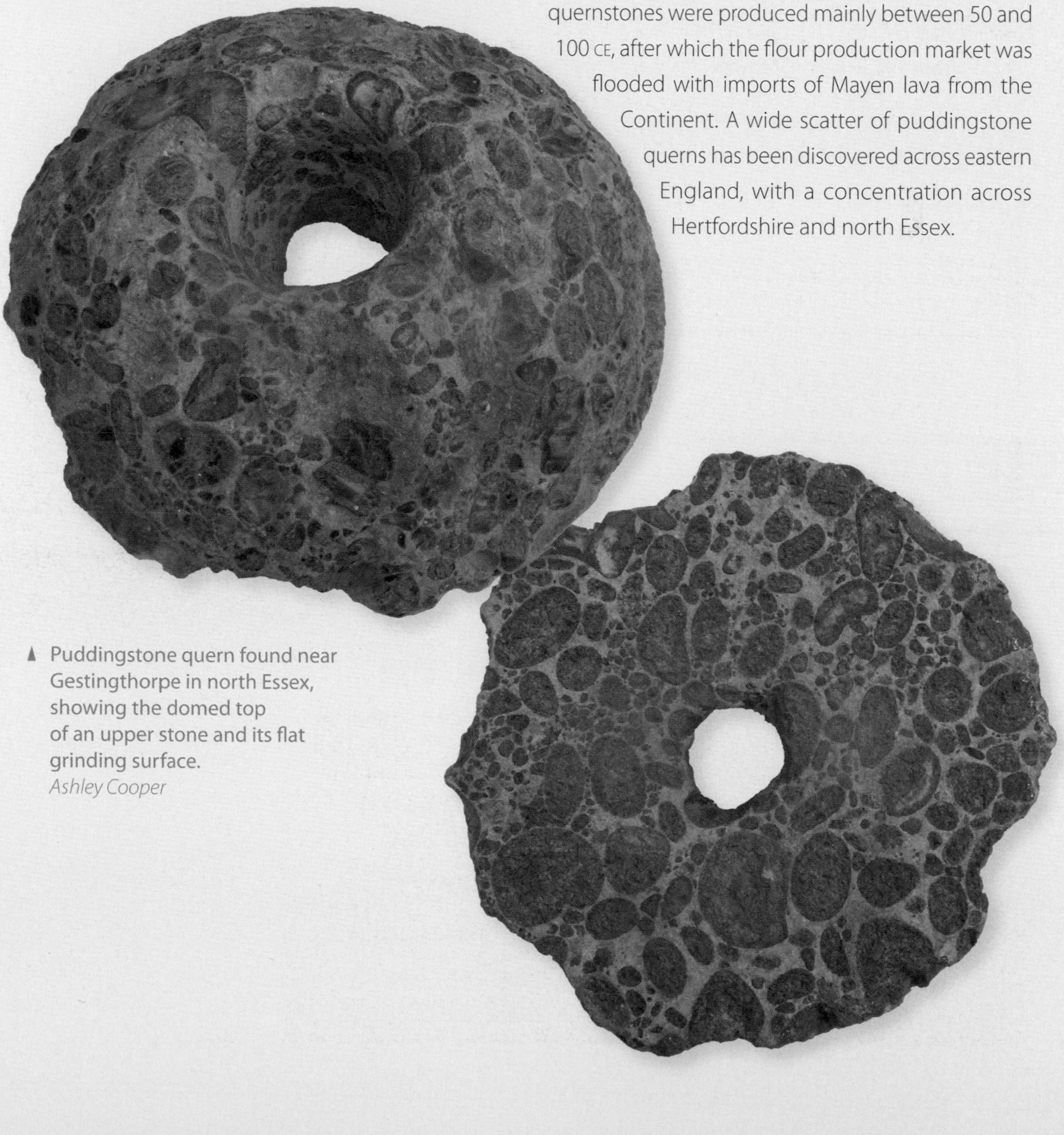

▲ Puddingstone quern found near Gestingthorpe in north Essex, showing the domed top of an upper stone and its flat grinding surface.
Ashley Cooper

Geology, soil and farming

Soil is the natural interface between geology and biology. It is a thin layer recycling dead stuff back into life. Throughout the many sudden episodes of cooling and warming during the ice age, the landscapes of Essex have been subject to repeated soil generation and destruction. Ancient soils have been detected on the surfaces of the early terraces of the Thames across north and mid-Essex – the Valley Farm Soil complex. These are mostly buried beneath later glacial till. The present-day surface soils of Essex are newer, formed after the last glacial episode, and they are varied, depending mainly upon the underlying geology.

Today's soils are largely derived from the forest soils that gradually developed during the warming after the last glaciation. This occurred around 15,000 years ago and again, after a cold snap, from around 11,700 years ago. The natural landscape of most of Essex in any interglacial stage is that of forest, mainly of deciduous broad-leaved trees such as birch and hazel, followed by elm, oak and lime. This present interglacial, the Holocene, is no exception. The fact that thick temperate forest, the original 'wildwood',

is almost nowhere to be seen in Essex is due to the action of its human populations over thousands of years.

Gradual tree clearance would have started very early, from small clearings about 11,000 years ago in Mesolithic times, and accelerated with the production of more efficient Neolithic flint tools, such as polished and hafted flint axes, between 6,000 and 4,000 years ago. After this time, the ages of metal usage enabled forest clearance to accelerate rapidly, albeit with sophisticated flint tools still very much in use. By late Roman times, it appears that large areas had been cleared east of London. Population growth through medieval times and beyond, despite setbacks from climate fluctuations and plagues, ensured the need for continued forest clearance as ever greater areas of land were brought under the plough.

Areas where woodland has remained, such as along the Epping Forest Ridge, show how the underlying deposits have influenced the nature of the soil and, in turn, the types of native tree. There, mixed hornbeam and oak woodland grows on the London Clay and beech on the sandier Claygate Beds. Elsewhere in Essex, tiny isolated remnants of wildwood remain, repeatedly reduced and under threat from human activity.

▼ An eroding prehistoric land surface near Stone Point, north of Walton. This soil layer overlies the ice age wind-blown loess layer and contains Neolithic and Bronze Age flint tools. *Bill George*

Soils take hundreds of years to develop fully in our temperate climate. Most soils across Essex contain some wind-storm silt and dust that spread across the landscape during the coldest parts of the last glaciation. These are the loess deposits that formed when easterly cyclonic storms blew dust across Essex from the Doggerland plains at a time when Essex was a deeply frozen, dry land devoid of vegetation. Most of Essex received at least a heavy dusting of this loess and some areas received and retained a deep cover. Loess contains a high proportion of calcium and this has made many soils in Essex more fertile than they would have been without this ice age dust storm input. The thicker loess and silty or sandy 'coverloam' soils of the Tendring plateau make fine grain-growing land, supporting a large local maltings and seeds industry.

The chalky boulder clay or till that covers much of north-west and central Essex supports soils that are a rich, crop-producing resource. The clay-rich soil is ready 'limed' with chalk, 'two-horse ploughland', making it less acid and more fertile and very suitable for arable crops.

The London Clay vale across from Romford to Maldon gives rise to less fertile soil, and its heavy nature has made arable farming more difficult, leading to small, dispersed settlements, and an emphasis on pasture. This 'three horse ploughland' was 'chalked'

▲ Beneath a surface covering of plants and wildlife, soil is forming from clay, sand and gravel, together with dead plants, small animals and microbial life. During this slow but vital recycling process, soil is built up by worms. An ancient soil shows as a black horizon across this picture. Thorndon Country Park near Brentwood.

▼ Comparing the pattern of geology beneath the soil with the Essex soil map reveals their relationship across the county. *Soil map: Essex County Council*

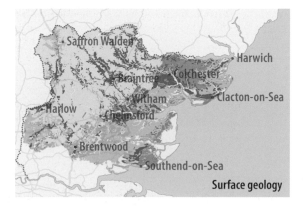

Surface geology

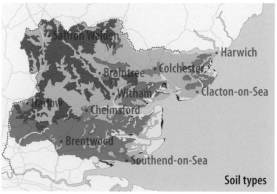

Soil types

every few years, by the application of lime to add calcium to break up the heavy soil. Lime was made from chalk dug from quarries or out of 'deneholes', ancient chalk mines, and heated until quicklime, calcium oxide, was formed. Quicklime reacts rapidly upon contact with moisture, combining with soil acids to neutralise them, thus improving the soil for agriculture. The chalky soil in the Purfleet area once supported vineyards, followed later by cherry orchards until the area was overrun by quarrying, followed by housing estates. In southern areas of the county where gravels from the Thames terraces lie at the surface, the soils are acid, barren and far less suitable for agriculture or horticulture. Some of these areas retain their original heath names, such as Chadwell Heath, Squirrels Heath and Daws Heath. The soil types of Essex have helped shape the landscape, wildlife and economy of the county. Essex soils are varied and include some of the best arable farming soils, which makes the county's farming crucial to our economy.

▲ London Clay soil in the foreground, with Claygate Beds soil beyond the stream, looking north from Buttsbury church, mid-Essex.

▲ High-calcium soil in north Essex near Ashdon, where no-till farming has produced a healthy soil with plenty of worms to plough the land.

The dynamic landscape

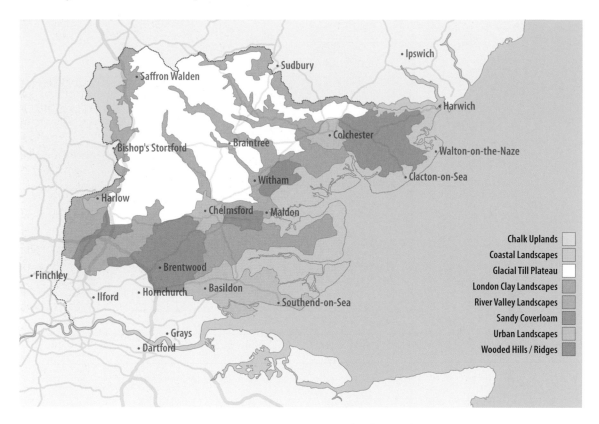

Chalk Uplands
Coastal Landscapes
Glacial Till Plateau
London Clay Landscapes
River Valley Landscapes
Sandy Coverloam
Urban Landscapes
Wooded Hills / Ridges

▲ Essex landscape character areas.
Map based on Essex County Council data

Alluvium

Alluvium is land being removed. You can watch Essex being taken away, sediment and soil, on its way to be recycled as layers of new sediment in the sea. On the scale of our own human lifetime, the constant erosion of the landscape of our county's surface rocks may not be immediately obvious. Yet the geological processes that shape the countryside are constantly active all around us. The alluvium sediment comprises soil, mud, silt, sand and gravel and organic matter, transported material that has been weathered and eroded from the lands of Essex. This is deposited and remobilised in stages along streams and riverbeds. During times of storm and flood, the rivers pick up large amounts of sediment, turning the waters muddy brown with their suspended load. These sediments eventually reach estuaries, marshes and the North Sea. For that is where Essex is going.

But Essex is still here. It has not been completely worn away with its every particle taken to the sea and re-deposited. It is still a land area, albeit with variably flooded edges. Yet the sea level has risen and is still rising. Likely near-future sea levels will flood greater areas of Essex. However, much will remain as land and this land will continue to be worn away. Overall, the lands of Essex have been rising slowly through millions of years, with smaller-scale sinking episodes taking place on shorter timescales.

These crustal movements are constantly active; they continue as ever. They seem insignificant on our human timescale, but their overall effect has been

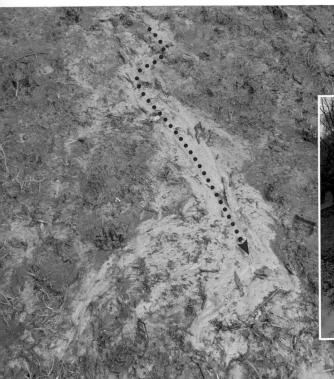

◄ Alluvium is moved by rainwater down a hillside at Hadleigh Country Park: the start of the journey of Essex rock to the North Sea.

▲ Muddy floodwaters: the River Chelmer carries alluvium on its journey to the North Sea.

to raise the north-west of Essex, tipping the crust down to the south-east. Consequently, most of the present area of the county has remained above the waves, and many of its streams and rivers flow from the north-west and the west, as do the Thames and the Stour along its borders. The North Sea area is sinking and acts as a 'basin', a stockpile of alluvially transported sediment from the lands of Essex. The seabed layers are a record of climate changes and sea level variations. As the land is eroded, its weight decreases and the crust rises a little, maintaining the land area; as the sediments accumulate across the North Sea bed their weight pushes the crust down a little, continuing the sinking of this sedimentary basin.

Across the gravels of the Thames terraces, there are patches of alluvium from rivers and streams mixed with wind-blown silt, the loess, plus sediments that washed down the slopes of the terraces. These are shown on geological maps as 'brickearth' and were indeed worked for brickmaking from Roman times. These slope-wash sediments may in part derive from soil disturbance by early agriculture. Deposits of alluvium from throughout the Holocene interglacial contain bones of mammals such as horse, elk, wild boar, bear and wolf, as well as submerged woodland, plus flint flakes and arrowheads made by humans, together with plenty of the much more recent debris from modern civilisation, such as plastic bags and supermarket trolleys. The study of the types of sediment and their relationships with each other, plus any flint tools, pottery and other objects trapped within the sediments, enables us to build up a picture of the environment and the succession of life, and of human culture and activity. The alluvium of the Essex coasts, rivers and estuaries is a rich store of evidence. As parts of the present coastline of Essex are being eaten away by erosion from the sea, other parts are being built up as new land. Such 'give and take' is better illustrated in Essex than anywhere else in Britain. As has been the story for so many millions of years, this county is edgeland – it is at the front line of change between sea and land. The present-day story of these coastal areas is an excellent illustration of what is happening now and what is likely to come.

Much of the Essex coastline is underlain by a geological layer that is soft and vulnerable to erosion: the London Clay – the 'soft-rock coastline' within the London Basin. This fine clay sediment is also prone to re-sedimentation where the clay minerals transported as river 'mud' are aggregated within the salt water of the North Sea, enabling them to settle and build up in quiet waters, creeks and flats, with biological assistance from plants, particularly adhesion of algal filaments.

Essex saltmarshes: biology, sediment and new rock

Saltmarshes – also known as saltings – are fragile and complex coastal belts, ever-changing areas where tide, sediment, plants, sea creatures, birds and time all combine to create a self-contained system without any human intervention. Essex contains 12% of all the saltings in the UK.

▼ Woodrolfe Creek saltings at Tollesbury. Walkways provide an unusual opportunity to observe these dangerous but vulnerable and valuable sediments.
Russell Wheeler

Saltmarshes and their fronting mudflats are among the few wilderness areas left in the UK. They are formed when silt and mud are deposited in sheltered locations and colonised by salt-tolerant plants. The surface of the marsh is dissected by a system of drainage creeks and often pitted with pools or 'salt pans'. Essex is especially important for this habitat – its coast has a larger area of saltmarsh than any other county. Woodrolfe Creek at Tollesbury on the River Blackwater is probably the best and safest place to see saltmarsh in Essex. Here the marsh is criss-crossed by paths to enable boat owners to reach their vessels. Radiocarbon dating of plant remains at a depth of 3.5 metres in a borehole at the edge of the saltmarsh at Tollesbury produced at date of almost 5,000 years. The growth rate of the saltmarsh has been nearly constant with an accretion of about 1.5 millimetres per year. This is new Essex rock in the making.

Saltmarshes form where the tide washes over the land twice each day; the fronting mudflats are formed in harmony with the saltmarsh, sediment moving from one to the other during on-shore gale attacks, so that the saltmarsh may build up further. Tidal currents can then form a maze of creeks. Because the marshes dissipate the energy of storm surges and wave action, their build-up in front of sea walls provides extremely cost-effective protection of the coastline. Around 300 km (200 miles) of Essex sea walls rely on a salting as the first line of defence against the tide. If these natural defences are lost, then resulting repair or wall strengthening costs would be immense. Saltings may redevelop naturally in shifted locations after wall loss.

While saltmarshes are susceptible to the threat of sea-level rise, they may keep pace given the right conditions. Marshes are subject to both erosion and build-up. As vegetation captures sediment, this allows the plants to grow better, trapping more sediment and accumulating more organic matter. This positive feedback allows the level of the saltmarsh bed to keep pace with rising sea level, unless prevented by human

disturbance. Ultimately, in a subsiding area or where sea level is rising over a long time interval, the marsh and mudflat sediments continue to accumulate, given a continuation of the supply of transported mineral matter and biological interaction. The sedimentary layers gradually become compacted. Water is expelled from the pore spaces and algal cells decay and add to the content of stagnant organic matter. Ultimately, compressed layers of shale or mudstone are formed at depth as burial proceeds over tens of thousands of years. New rock layers begin to form. A few trapped plant and animal remains may eventually become fossilised if the new layers become more deeply buried and if they are not eroded away, such as during a major storm.

Managed breaches and reserves

Around 90% of coastal marshes in the UK have been lost in the last 400 years and others are disappearing because of rising sea levels. The huge North Sea surge of December 2013 was one occasion when Essex farmland was inundated, and all the indications are that similar extreme events will reoccur thanks to climate change and sea-level rise. In certain parts of the Essex coast, breaches of the coastal barriers have been managed in order to allow a return of the area to a more natural habitat for wildlife. For instance, in 2015, a 300 m stretch of the sea wall was breached at Fingringhoe, to let the sea flood into a 22-hectare area and re-form the intertidal habitat of saltmarsh, mudflats, saline lagoon and reed beds. Now, twice daily, the tides of the Colne estuary cover and expose Fingringhoe's new intertidal mud area.

The area of Fingringhoe Wick Reserve has a long history, with evidence of Roman occupation. For many years the area was farmed before being sold for sand and gravel extraction. From the early 1900s to the end of the 1950s the site was a busy industrial area. The sand and gravel left by the last course of the River Thames across mid-Essex and reworked by glacial meltwater, the Lower and Upper St Osyth Gravels, were quarried and taken by river to London. The geological significance of such areas is generally ignored, but their story is an important one of interaction and change; monitoring the relationships of sediments and biology in such areas is just one aspect of understanding the past, present and future of coastal changes and the effects of climate variation.

◀ The 2015 breach at Fingringhoe creates a new intertidal habitat across a former industrial gravel extraction area. *Environment Agency*

Foulness Island

Remote and isolated, Foulness (formerly called 'Fowlness') is the largest of the Essex islands, over 2,400 hectares (6,000 acres) of flat, almost treeless land. At Foulness Point, saltmarsh is fringed by a prominent bank of cockle shells known as a 'chenier ridge' similar to that a little further north at Bradwell-on-Sea. These masses of empty shells have been eroded from tidal flats and built into ridges by waves and storms. They slowly migrate landwards until they are stabilised by saltmarsh vegetation. They form the most extensive shell beach in Britain. Access to Foulness Island is restricted to those that live and work there as it is used as an artillery range by the Ministry of Defence.

Until a bridge was built in 1922 the main access to the island was along the 'Broomway', recorded from 1419, but probably established in Roman times. It was formerly marked by markers resembling upturned besom brooms and has long been notorious as the most perilous byway in England. Travellers become disorientated by the poor visibility and may be caught by the tide.

▼ Exploring a land of new rock: a Geologists' Association visit, travelling to Foulness in 1911 along the 'Broomway', a causeway built along the fronting mudflats of the Foulness saltings.
The Geologists' Association archive

Landslips

Landslips, landslides, cliff falls and slow soil creep are all phenomena that can result from the action of gravity on rocks, sediments and soil wherever there is a slope or an edge. With rain, freezing and thawing, the percolation of water through joints and spaces between grains of sediment in the ground, flooding and drying out, there is an eventual tendency for rock layers to fail or shift, either suddenly or gradually. The

London Clay is particularly susceptible to changes in water saturation, becoming plastic and tending to flow when wet and to shrink, harden and crack when drying out.

Landslips occur along many stretches of the Essex coast and along river estuaries where steep slopes have been formed in the London Clay. These landslips probably originated during periods of periglacial activity with successive cycles of freezing and thawing. Slipping on various scales has been taking place for thousands of years and will continue until an angle of about 8° is reached for ultimate stability on London Clay slopes.

Successive rotational slips occur especially where London Clay is overlain by permeable deposits such as the Bagshot Sand and Claygate Beds or ice age sand and gravel. These beds feed water into the London Clay over large areas. In some places there has been considerable investment in preventative measures. Coastal defence structures protect the cliffs from erosion at the Naze north of Walton, deep piling into London Clay prevents movement at Holland-on-Sea, and construction of sea walls and drainage of the London Clay at Southend slows the rate of slumping. There are sites of known instability that have constrained development in some areas. Ultimately nature will prevail as the landscape continues to evolve. Examples of significant landslips are described in the following pages.

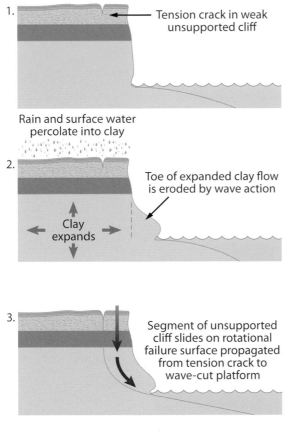

1. Tension crack in weak unsupported cliff

Rain and surface water percolate into clay

2. Toe of expanded clay flow is eroded by wave action

Clay expands

3. Segment of unsupported cliff slides on rotational failure surface propagated from tension crack to wave-cut platform

4. Toe of rotated section further eroded by direct wave action and mudflows

◄ When water seeps into cracks in London Clay it is absorbed, causing the clay to expand. The outward pressure leads to ground failure and landslips develop. Landslips are common throughout Essex, even on gentle slopes inland.

Pitsea Mount: St Michael's Viewpoint

Landslips are common between here and One Tree Hill to the west, due to the steep slopes that reveal an abandoned River Thames cliff line. St Michael's Hill consists of London Clay with slopes of 9 to 10°, too steep to ensure stability, which may be the reason behind the abandonment of the church. All around the hill are cracked paths as the ground continues to move. It may take thousands of years for the hill to attain a stable slope.

Hadleigh Castle landslip

Study of the clay layers mantling the cliff has revealed several major phases of landslipping. The first occurred under periglacial conditions shortly after the cliff was abandoned by the River Thames. Temperate slipping occurred from 7,000 to 6,500 and from 2,100 to 2,000 years ago in response to varying climate conditions. Most recently, there has been a phase that started with a major rotational slip in Victorian times, around 1890. The degraded part of the cliff has an average inclination of over 12° and is still actively landslipping. This indicates that the total time required after abandonment for the cliff to reach the ultimate angle of stability, about 8°, could be several times that which has elapsed so far. Open tension cracks in the ground are evidence of continued movement.

◄ Landslip at Hadleigh Castle.
Essex County Council

Southend-on-Sea

There is a large rotational slip between Hadleigh Castle and Southend-on-Sea. The footpath follows the up-tilted edge of the slip on the Bagshot Sand which overlies the London Clay here. Towards Southend-on-Sea, the coastline takes the form of a line of cliffs up to 30 metres (100 feet) high of London Clay, overlain by ice age sand and gravel and river alluvium. Before the development of Southend in the nineteenth century this coastline was under constant attack by the sea and would have consisted of tumbled and vegetated slopes with mud flows and large masses of London Clay moving seawards along slip planes lubricated by groundwater. Today, the movement of the ground along much of the Southend coastline is a constant problem for the Council engineers, the most serious landslips occurring at The Cliffs, a section of coast just over a kilometre long (approximately three-quarters of a mile) between the Cliffs Pavilion and Southend Pier. Since the sea wall and promenade were built in 1902 the cliffs have mostly been laid out as public gardens but this has not prevented the landslipping. The cliffs here are fairly steep (up to 30° in places) and despite extensive attempts at underground drainage, the ground frequently creeps forward which results in cracked paths and leaning lamp posts and trees. Occasionally there are major landslips such as the one that occurred in December 2002, which was 180 metres (600 feet) wide and destroyed large areas of the gardens, threatened the bandstand and temporarily closed two lanes of the seafront dual carriageway.

▲ Looking towards Southend from Hadleigh Castle – a rotational slip of London Clay. The grassy area to the right has slipped from the left, leaving a valley. The slipping mass of clay rotated from the junction with the overlying Claygate Beds (the upper part of the field at the far left) to form the ridge on the right topped by the straight footpath.

Bushy Hill, South Woodham Ferrers

Bushy Hill is a ridge of high ground overlooking South Woodham Ferrers, which is an old river cliff of the River Crouch. Landslips of several different types have been taking place for thousands of years and will continue until the slopes reach their angle of stability.

The slippage of the London Clay is exacerbated by springs emerging from seams of sand in the Claygate Beds above. The hummocky ground with scars from previous landslips can be seen on the slopes to the south and west.

West Maldon landslip

There are three landslips on the north-facing river cliff of the Blackwater at Maldon. The middle and most obvious slip is referred to as the West Maldon Landslip. It is formed of successive rotational slips in the London Clay which is very gradually attaining a stable angle. Most of the slips are grassed over but small fresh scars indicate that there is still instability. The most active landslipping presumably took place when the river

was directly eroding the toes of the landslips. The presence of saltings indicates that erosion has not taken place for a number of years. The surface of the landslipped ground is characteristically hummocky but now trees have mostly covered the slope. As sea level rises and overwhelms the saltings, then the landslips will resume at a faster pace.

Jupes Hill landslip

A landslip in the London Clay slopes on the south side of the Stour valley near Dedham is reported to have occurred in 1928. An area of about 80 m (260 ft) across the slope and 80 m (260 ft) down, on a slope of between 9° and 11.5° between the 30 m (100 ft) and 20 m (65ft) contours was involved in the movement which left a rear scarp of about 2.5 m (8 ft) in height. The toe of the slope has now been ploughed away. Little evidence of the landslip can now be seen, but there is evidence of larger, older landslips along this side of the valley.

Creep

Slow processes of 'soil creep' have affected the landscape throughout the ice age and continue today. The surface layers of the landscape are redistributed, flattening its slopes. On slopes where there is London Clay at the surface, soil-creep can transport sediment down very shallow gradients. The clay minerals absorb water readily and, in doing so, expand to heave the surface upwards. Upon drying and contracting, the clays slump back down, edging downhill. Vegetation in the soil layer tends to consolidate the clay into layers to form terracettes that run along slope. The curved trunks of trees also show that this process is still going on. Thus, the impermanent nature of the surface layers of a hill slope is revealed in the landscape. The inherent instability of slopes in the London Clay needs to be an important consideration in the design of major engineering projects such as road and railway cuttings and embankments.

▲ Soil-creep slip terracettes in a London Clay slope at Thorndon Country Park.

'Tree creep' evidence: tree growth ►
changes during soil creep.

Disappearing Naze Hill

The spectacular erosion of the tallest cliffs at Walton-on-the-Naze is caused by rotational landslips. Cliffs are inherently unstable because their outer side is unsupported, but here the presence of London Clay is an additional factor. Water penetrating the permeable upper layers of the cliffs feeds into the London Clay beneath, causing it to swell. The weight of the overlying sediments squeezes the Clay, which then flows outwards at the base of the cliffs forming a toe which is then eroded by wave action. This sets up instability within the cliff itself which fails along a curved surface. Repeated landslides have eroded the cliffs by a considerable distance over historical time. Scars of previous rotational slips are marked by curved ridges on the foreshore, as seen at low tide. At the north end of the Naze, where the cliffs are much lower, the erosion is caused more by direct wave action but has, nevertheless, been significant. The 110 m (360 ft) Crag Walk, completed in 2011, was built along the Naze shore to help protect the Site of Special Scientific Interest (SSSI) and Naze Tower from the sea. More than 16,000 tonnes of larvikite rock were shipped from Norway to build the walk, which is expected to slow the erosion to 20 m (65 ft) over the next 70 to 100 years, as opposed to the previous rate of erosion of 1 to 2 metres each year. This also means that this nationally important geological SSSI site is being obscured rapidly and made inaccessible by rampant vegetation – something of a dilemma.

▲ Cliff cracks and slumps at the Naze.
A grass-topped rotated mass has slipped down (centre left); vertical cracks in the foreground show where the next mass is ready to slip.

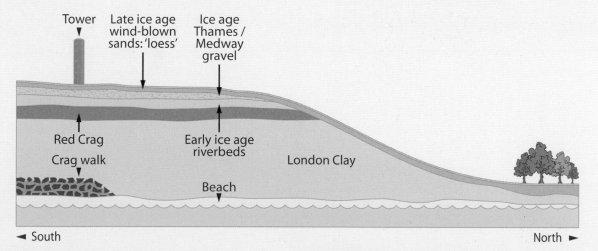

◄ South

North ►

▲ The rock layers making up the Naze Hill.

The remnants of the ► 'Tamarisk Wall' at the north end of the Naze beach can still be spotted at low tide. This early twentieth century prom and sea defence was finally wrecked during the 1953 flood, after which parts were reused for further defences.

Walton and the Naze – erosion 1300 to 2020. ► The coastline was said to be 'a quarter of a mile from the Tower when it was built in 1720'. The old All Saints church in Walton was inundated by the sea by 1789, its remains finally being washed away by the tide during 1796, with further houses and shops carried away by the sea by 1820. The direction of Kirby Road and the High Street in Walton town, which now stops at the seafront traffic lights, indicates the former way through the lost village.

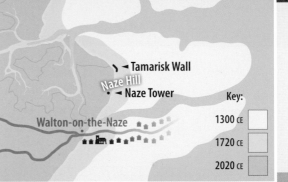

Tamarisk Wall
Naze Hill
Naze Tower
Walton-on-the-Naze

Key:
1300 CE
1720 CE
2020 CE

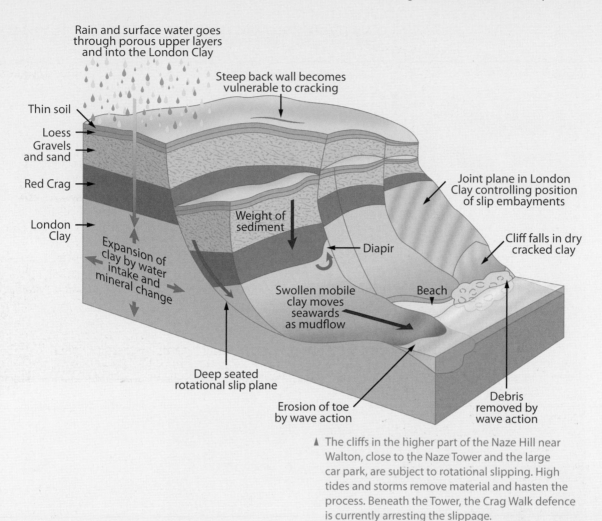

Rain and surface water goes through porous upper layers and into the London Clay

Steep back wall becomes vulnerable to cracking

Thin soil

Loess

Gravels and sand

Red Crag

London Clay

Expansion of clay by water intake and mineral change

Weight of sediment

Joint plane in London Clay controlling position of slip embayments

Diapir

Cliff falls in dry cracked clay

Swollen mobile clay moves seawards as mudflow

Beach

Deep seated rotational slip plane

Erosion of toe by wave action

Debris removed by wave action

▲ The cliffs in the higher part of the Naze Hill near Walton, close to the Naze Tower and the large car park, are subject to rotational slipping. High tides and storms remove material and hasten the process. Beneath the Tower, the Crag Walk defence is currently arresting the slippage.

▼ The rapidly eroding Naze Hill from the low tideline.

◄ Erosion of Naze Hill by landslipping. The Naze Tower and cliff line in 1977.
Reproduced with permission of the Cambridge University Collection of Aerial Photography, copyright reserved

◄ A staircase of slipped segments below the Tower in April 1997.
Essex County Council

◄ The newly built Naze Crag Walk below the cliffs and Tower in 2011. This sea defence is delaying erosion of the cliff directly behind it.
Breheny

A circular 'sole scar' of a previous rotational cliff ▲ slippage imprinted in the eroded platform of London Clay on the Naze beach.

Concrete gun emplacement ► at Walton-on-the-Naze beach in 1966.
Bob Markham

The same structure in 2018, showing its increased ► distance from the receding cliff foot after 52 years.

Swallow holes

In north Essex, north-east of Gestingthorpe within the acreage of Hill Farm, a feather edge of sand and gravel, the Lower London Tertiaries, lies above the Chalk where a valley has been incised by river erosion. Topping all of these beds is the layer of Kesgrave Sand and Gravel laid down by the ice age River Thames, followed above by the edge of a layer of glacial till, with a variety of erratic rocks, left by the Anglian ice sheet.

Rainwater percolates through the river gravel of the Kesgrave Sand and Gravel, and then down through the Lower London Tertiaries until it meets the Chalk below. The fissured Chalk dissolves gradually in the acid water, widening out the cracks until an opening in the chalk becomes wide enough to 'swallow' the gravels, which cave in above the hole.

Across the ploughland, swallow holes have suddenly appeared, especially when a heavy horse or a tractor was at work. Several instances of sudden collapse are recorded. Farmer Ashley Cooper has written a graphic account from farmworker conversations in his book, *The Long Furrow*:

Overidges Field grew all twitchy so we ploughed it up and used three horses on the plough so that we could get hold of it in a bit deeper … Well, we come to pull about, we took the drag harrows with four horses … When we got to the narrowest bit of the field, one of the horses suddenly stopped. It had got its hind hoof through the bottom of the furrow and couldn't get it out … we went to the front and undid all the leaders and lines but … the horse was gradually sinking into this hole … come to the finish it was almost disappeared altogether … so we got some spades and cut a trench in front … and got it out that way. But this old horse never really got over it, and my father who had worked on the farm all his life was hully upset as well.

◄ A heavy horse is rescued after being trapped in a subsiding swallow hole on Hill Farm, Gestingthorpe. Drawn from recollections of a farmworker from 1921. The poor horse never recovered from the frightening ordeal. *Ashley Cooper*

▲ Farmer Harold Cooper stands in a swallow that appeared on his land in 2000. *Ashley Cooper*

Where streams flow nearby, they can disappear into swallow holes if the percolating water erodes through the stream-bed sediment (alluvium). The water dissolves and cuts down to the Chalk beneath, leaving the stream bed dry further down its original course.

▼ Swallow holes form in and around the feather-edge of rock layers above the Chalk near Gestingthorpe in north Essex. Water drains off the land and through the sand and gravel layers. Water seeps down into the Chalk below, widening fissures until the land above collapses into the cavity beneath. Stream water flowing across the Chalk disappears down holes in the stream bed.

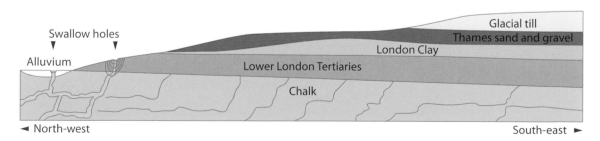

Glacial till
Thames sand and gravel
London Clay
Lower London Tertiaries
Chalk
Swallow holes
Alluvium

◄ North-west   South-east ►

Heave damage from shrink and swell in clay

Heave damage, from shrink-swell action through dry and wet spells, has been a major hazard to building and construction in Essex, particularly in the areas of London Clay, but also in till ('boulder clay') areas. The clay minerals in the rock and overlying soil include the type in which each minute clay-mineral crystal swells upon contact with water, drawing it in to become part of its crystal structure. The whole mass of clay thus swells up as it becomes wet. This is called 'heave'. Upon drying out again, the clay mineral loses its water and contracts, and the whole mass shrinks causing

▲ Cracks showing where the right-hand side of the wall has been heaved upwards by rehydrated clay swelling beneath a house built on London Clay at Colchester.

the ground to crack open. Water trickles in again and the land swells, often rising beyond its original level. This upward heave tends to push anything built upon or into the clay and can cause considerable damage. The removal of trees is a major cause of heave. Trees are extremely efficient in their ability to follow water and extract it from rock and soil, often at remarkable distances of several tree diameters from their trunks. When a tree is cut down, water is no longer taken up by the roots and remains in the soil, causing the clay to expand beneath a building and cause damage by heave. This is a far more common problem than that caused by clay shrinkage, although this may have an effect in exceptionally dry periods.

Insurance claims over many years correlate well with areas underlain by London Clay. Although building methods now include rafts and piles to combat damage due to shrinking and swelling of the clay, trees and their removal still cause local problems in clay areas.

Subsidence tends to occur wherever a cover of unconsolidated sediments or weakened built ground is undermined by gradual ground movements or solution. Failure of ground above a pre-existing but unsuspected cavity, such as a forgotten and unrecorded well, can take people by surprise.

An old well ► is rediscovered near Coval Lane, Chelmsford.
Gerald Lucy

Earthquakes

Earthquakes are rapid events that can be all too apparent to humans. However, they are evidence of very gradual changes in the Earth's crust and, consequently, long-term alterations in the geography of the lands above. Continual adjustments in the basement rocks of the Anglo-Brabant Massif, the thick crust that extends beneath Essex and across to Belgium, are caused by long-term tectonic actions, notably the drift of the African continent towards Europe. Eventually, these movements affect the landscape but, in the short term, the earth literally quakes, even if only slightly.

Although it is generally thought that earthquakes are rare in Britain, they do occur very frequently on a geological timescale. Approximately 300 are detected each year by sophisticated monitoring equipment and of these about 30 are strong enough to be felt. Occasionally, however, Britain is shaken by an earthquake that causes structural damage.

Seismograms from a ▶ small earthquake near Harwich, 15 September 1994, automatically recorded at Folkestone (top three traces), Eastbourne and Brentwood. The epicentre was 40 km offshore and measured 3.2 on the Richter scale.
British Geological Survey

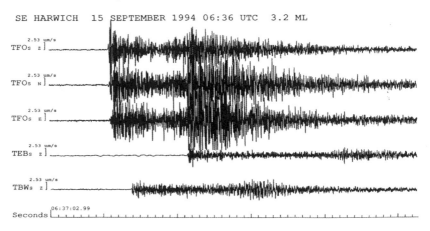

The Brabant Earthquake 1692

There is evidence of a large earthquake in the parish records of St Peter's Church on North Hill, Colchester. The earthquake occurred on 8 September 1692 and the register records the damage to the central tower of the church:

On Thursday, Sept. 8, 1692, there happened about two of the clock in the afternoon, for the space of a minute or more, an universal earthquake all over England, France, Holland, and some parts of Germany. And particularly it was attested to me by the masons that were there a-plastering the Steple of St Peter's, in this town, and upon the uppermost scaffold, that the Steple parted so wide in the midst that they could have put their hand into the crack or cleft, and immediately shut up close again, without any damage to the workmen (who expected all would have fallen down), or to the Steple itself. Most of the houses here and elsewhere shook, and part of a chimney fell down on North Hill; and very many who were sensible of it were taken at the same time with a giddiness in their head for some short time. In witness of what is here related, I have hereto set my hand, Robert Dickman, Minister of St. Peter, Colchester.

The British Geological Survey's Catalogue of British Earthquakes (Musson 1994) confirms that the earthquake, known as the Brabant Earthquake, had an epicentre within the underlying crust of the Anglo-Brabant Massif between Brussels and Liege and was widely felt in the Low Countries, north-east France, Germany and south-east England.

The Colchester Earthquake 1884

The most destructive earthquake ever recorded in Britain occurred in Essex on the morning of 22 April 1884 and strongly shook most of the county. It is known as the Colchester Earthquake because the greatest damage was caused to Colchester and nearby towns and villages such as Wivenhoe. The tremor was felt over much of southern England and parts of France and Belgium, and its magnitude has been estimated at 5.2 on the Richter scale.

The number of casualties is difficult to estimate, but it is doubtful whether any deaths or serious injuries can be attributed to the earthquake. There was, however, considerable damage to over 1,200 buildings in Essex. The earthquake was due to movement along a fault in the basement rock of the Anglo-Brabant Massif which shook the overlying layers of Cretaceous and later rocks. An extensive study of the effects of the earthquake was carried out shortly afterwards by the Essex Field Club and their detailed report, published in 1885, is a fascinating and extremely valuable account.

▼ A front-page drawing from the *Illustrated Police News*, May 1884, showing Lion Walk Congregational Church. The repaired section of the steeple top is discernible today.

▲ Ann Watkins, the owner of Wayside Cottage, Church Road, Peldon, was so proud of its history that she had a pottery plaque made for the outside wall of the cottage. *Peldon History Project*

EFFECTS OF THE LATE EARTHQUAKE IN ESSEX.

CHIMNEY OF PUBLIC-HOUSE, PELDON ROSE.

BROKEN TOMBSTONE, PELDON CHURCHYARD.

▲ Earthquake damage to a house at Great Wigborough. *Essex County Libraries*

▼ Structural damage map of the 1884 Colchester Earthquake.

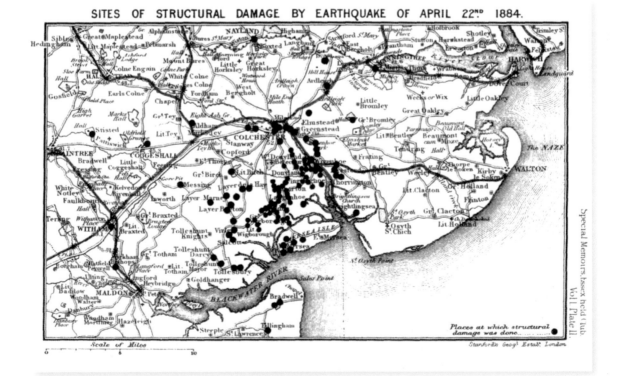

SITES OF STRUCTURAL DAMAGE BY EARTHQUAKE OF APRIL 22ND 1884.

The Ashdon meteorite

On 9 March 1923, a meteorite fell at Ashdon in north-west Essex. The fall was witnessed by a farmworker. This is a rare example of a meteorite fall being witnessed and the stone recovered.

Frederick Pratt, a thatcher, was working in the corner of a wheat field on Ashdon Hall Farm when, at about 1 p.m., he heard a strange hissing sound in the sky and looked up, supposing it to be an aeroplane. A second or two later he saw an object fall about 30 yards from him into the field causing the earth to 'fly up like water'. Three days later, in the company of another worker on the farm, he dug up a stone from a depth of about 60 centimetres (2 feet) and took it to the police station. He subsequently passed it to the vicar of Wendens Ambo who fortunately donated it to the Natural History Museum in London. Scientific investigation proved beyond doubt that the stone was a meteorite. This was an object that had travelled billions of miles through space and was at least ten times older than any other stone to be found in the soil of Essex.

From Pratt's observations of the direction from which the sound came and the inclination of the hole, it would seem that the stone was travelling south-west to north-east and must have passed over Saffron Walden. As far as could be ascertained there were no reports of sonic booms or detonations which would have been expected as the stone travelled at supersonic speed. During its descent through the atmosphere frictional heating would have turned it into a fireball. Had Pratt been able to handle the stone immediately after impact he would have found it warm to the touch but the interior would still have been exceptionally cold – the temperature of deep space.

The Ashdon meteorite is important not only because it is well-preserved and still available for study but because the fall was witnessed. In England in the last 100 years only three other meteorites were seen to fall and were subsequently recovered. The meteorite is in the Natural History Museum in London and, although it is not currently on display, it can be seen by appointment. A cast of the meteorite is on display in Saffron Walden Museum. Ashdon Village Museum also has a cast of the meteorite and a small display on its fall.

2 cm

◄ The Ashdon meteorite. As the surface of the stone melted during its flight the red-hot molten rock was forced backwards to be lost as a cascade of droplets to the air. As the stone slowed to subsonic velocity the final melt solidified as radial ridges and grooves.
Saffron Walden Museum

Stories in the Essex landscape: looking back into time

If we cycle or walk in Essex, the hills seem far steeper than if we travel by car; but we can stop and rest, and perhaps take a look. To pry into the landscape and look back in time, we could just consider why the valley is there and why its sides happen to be so steep. A look at the geological map reveals 'fingers' of drainage that cut valleys, some of them steep-sided, into the flat or undulating scenery of the glacial till plateau spread across much of Essex. Many of these valleys contain a bed of alluvium, shown on maps in pale cream colour. The alluvium may cover a very flat strip in the bottom of the valley, much wider than the river or stream that flows along the valley. This flat part is the floodplain and there are buildings or homes on many of them. That seems peculiar to a geological time-traveller, as such a stretch of country is really a plain that happens to flood, hence its name. Geologically, such floods

◄ The view across the Pant/Blackwater valley near Bocking, north of Braintree.
The houses lie upon glacial till. Thames gravels underlie the till, midway across the far side of the valley. A spring line lies above the London Clay closer to the valley floor and its alluvial floodplain.

☐ Alluvium
☐ Glacial till
■ Thames gravels
▨ London Clay

◄ Geological map showing alluvial valleys around and through Braintree. Erosion has cut into the top layer of glacial till and down through the Thames gravels into London Clay beneath.
Recent alluvium along the river valleys of the Blackwater and Brain. The view across the river valley is seen in the photograph.

River Pant

Viewpoint •

• River Blackwater

• Braintree

• Rayne

• River Brain

• Great Notley

• River Ter

• Silver End

are so frequent that the floodplain can effectively be considered as the river. So, geologically, these buildings are in rivers. This is also true for a large part of east London and south Essex, where housing and industry are built within what is effectively the River Thames and its tributaries.

The alluvium consists of the lands of Essex being removed and transported towards the sea. Erosion of the land continues. Soil, surface layers of glacial till and, eventually, sand and gravel from beneath the till, are steadily removed by rain and floods, to be trickled or swept downhill until they form part of the alluvial flows and floodplains along the valleys. The gravels tend to resist erosion rather more than the softer London Clay beneath. The clay is undermined, and may slip, creating steeper slopes as more gravel slumps downwards. The valley sides are steepened and the floodplains are fed with more alluvium. It is a continual process and it has not stopped even if we choose not to see it. It happens very quickly in areas where land is artificially exposed by development works or ploughing, situations that are unnatural and that aid erosion.

Land of the Fanns: a changing ice age landscape revealed

The area of fenland and low hills surrounding Bulphan village, the 'Land of the Fanns', is a rare and unusual example of how underlying geology profoundly influences the present landscape and its human interactions. Here, a fen basin lies in a bowl-like area with a single water outlet draining around a low chalk ridge. It is also a graphic story of how the land has changed with time, and will continue to change – whether or not we are here to see it.

The hills to the north and east of the bowl-like fen area are capped by 'High-level' gravels that were deposited by rivers that flowed northward across the area between half a million and 3 million years ago. Surprisingly, these flint gravel riverbeds now top the

Topography of the ►
Land of the Fanns.
*Land of the Fanns
Partnership*

highest areas of south Essex, following gradual uplift of the land by more than 100 m. The gravel-topped hills of Warley, Brentwood and Langdon Hills are prominent examples where the resistant gravel has protected the underlying soft sediments. Even the low hills followed by the road through South and North Ockendon may

▲ The gravel-topped Langdon Hills beyond the fen, viewed from Fen Lane.

be the eroded-down remnants of previously gravel-topped uplands. The unprotected surrounding clay areas were more vulnerable to erosion and have been worn away more quickly.

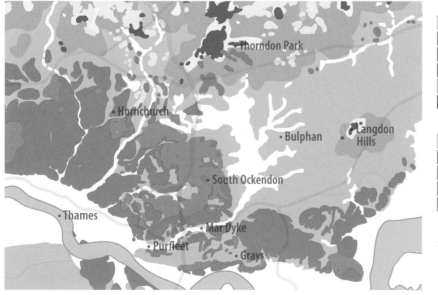

Superficial layers:

☐ Alluvium
▨ Brickearth/coversand
▨ Post-Anglian river deposits
▨ Anglian: glacial outwash
☐ Anglian: glacial till
■ Pre-Anglian river deposits

Bedrock:

☐ Bagshot Sand
▨ Claygate Beds
▨ London Clay
▨ Lower London Tertiaries
▨ Chalk

◀ The geology forms the foundations of the hills and fenland basin in this unusual landscape.

When the Anglian ice sheet extended south 450,000 years ago, and stopped locally at the Brentwood-Shenfield ridge, the ice front occasionally produced torrents of glacial outwash from spring snowmelt. Large spreads of outwash sediment were recycled from the ancient High-level Gravels of the Warley and Brentwood hills as the icy waters flowed away to the south. The waters with their sand and gravel load joined the huge River Thames as it swept across the areas now occupied by Ockendon, Grays and Stanford-le-Hope, probably scouring out a swathe of clay on the way. Some of this outwash sand and gravel is still to be seen in the lowest bed of the section at Thorndon Country Park.

During the final stages of the following glaciation, 320,000 years ago, snowmelt torrents across the permafrost tundra would have flowed down across the area of Thorndon Country Park, then devoid of vegetation. The waters gradually eroded the previous outwash gravels and the underlying sandy clay sediment, the Claygate Beds, eventually cutting right down into the London Clay beneath. The land had risen over the previous 100,000 years and deep valleys were carved into the soft sediments, lining them with gravel and flowing out across the flat clay area around Bulphan. Unable to escape through the rising ground of the surrounding gravel and clay hills, the trapped waters and clay sediment created a ponded-up 'bowl' which subsequently formed an area of fenland within the confining circle of hills. The icy water eventually found its way west around the chalk ridge of Purfleet by following the old Thames course. It incised a wide, flat-bottomed 'wadi'-like valley: the forerunner of the Mar Dyke.

▲ The dark gravel layer at the base of the ice age wall at Thorndon Country Park was left after huge floods came from the edge of an ice sheet at Brentwood 450,000 years ago.

The pebbles and sand ► of the glacial outwash gravel bed.

◄ Looking across a steep-sided valley in Thorndon Country Park. Here, ice age floods have cut through the gravel surface and the Claygate Beds, right down into the London Clay beneath. The icy waters drained into the fenland to the south, and on into the Thames.

◄ Looking south across the 'Land of the Fanns' from Thorndon south. The low Chalk ridge in front of the large pylon causes the Mar Dyke to flow west out of the fenland. Across the far horizon are the Chalk hills of North Downs in Kent. The Purfleet ridge is an outlier of Chalk – geologically part of Kent in Essex.

The present Mar Dyke stream occupies a flat valley previously scoured by icy floods during the last glacial period. The stream flows out from the Bulphan fenland which has now been drained for use as farmland. A future sea-level rise of a few metres would flood the fen area through the Mar Dyke valley to form a circular embayment. This fascinating and somewhat forgotten area of southern Essex provides a wonderful story in which geology and geography reveal how past climates have left a strange landscape hidden in plain sight.

A grand day out in Purfleet: a deeper look through time and landscape

Driving south along the M25 in Essex, it is hard to believe that a couple of kilometres before reaching the QE2 Bridge at the Dartford Crossing you are cutting across the bed of an ancient River Thames hundreds of thousands of years old. The approach road gradually rises from the old riverbed to take you over the Purfleet chalk ridge, now largely scooped out with old chalk quarries, including the quarry now occupied by the Lakeside Shopping Centre. The motorway here is perched atop a narrow wall of solid chalk between two large quarries. In Greenlands Quarry, a short way to the west, a partly overgrown corner is now part of an internationally important SSSI – kept under lock and key. Above a flat standing area looms a quarry face in the corner, revealing several layers of sand and gravel. These layers are banked against a mound of broken chalk; the contrast of golden sand and white chalk is surprising at first sight. This chalk is, in fact, the south bank of that ancient Thames. The sandy layers are the succession of riverbeds that happened to survive erosion long enough to be buried in the landscape after the Thames departed from its course along the edge of this low chalk ridge.

Down here, on the ancient riverbank, the chalk was exposed to glacial temperatures before the icy summer torrents washed into the bank and froze each winter. The chalk suffered freezing and thawing, the ground ice breaking up the soft rock into jagged lumps. Big, weirdly shaped pieces of flint dot the chalk, their broken surfaces showing up black and glossy in the powdery white slope. Adjacent layers of river sand and gravel reveal shelly fossils, flint pebbles and, in more than one level, pieces of flint that have been shaped or flaked by human beings. Not modern humans like us, but humans of another, ancient species. Those people were expert hunters. They had to make their work-tools quickly and efficiently and had to be able to predict and plan ahead. They survived in communities that visited here for hundreds if not thousands of years, who came and went with changes in climate over tens of millennia. Here, in Purfleet, was a prime hunting area by a big double-bend of the ancient Thames. This was their home beach, their look-out and their game territory. They lived just here, more than 300,000 years ago.

◀ The Mar Dyke is now just a trickle along the centre of a wide, flat-bottomed valley with steep walls, typical of a flood outwash valley.

Early Neanderthal humans occupied a prime ▶ hunting area in a loop of an ancient Thames river. Around 300,000 years ago this chalk ridge at Purfleet provided an ideal look-out. It was also an accessible site for extracting and making flint tools for butchery, spear making and many other uses. In the mid-distance the Mar Dyke flows out to the Thames; beyond, the ground rises to the long gravel ridges of Brentwood, Thorndon and Langdon.

Peter David Scott

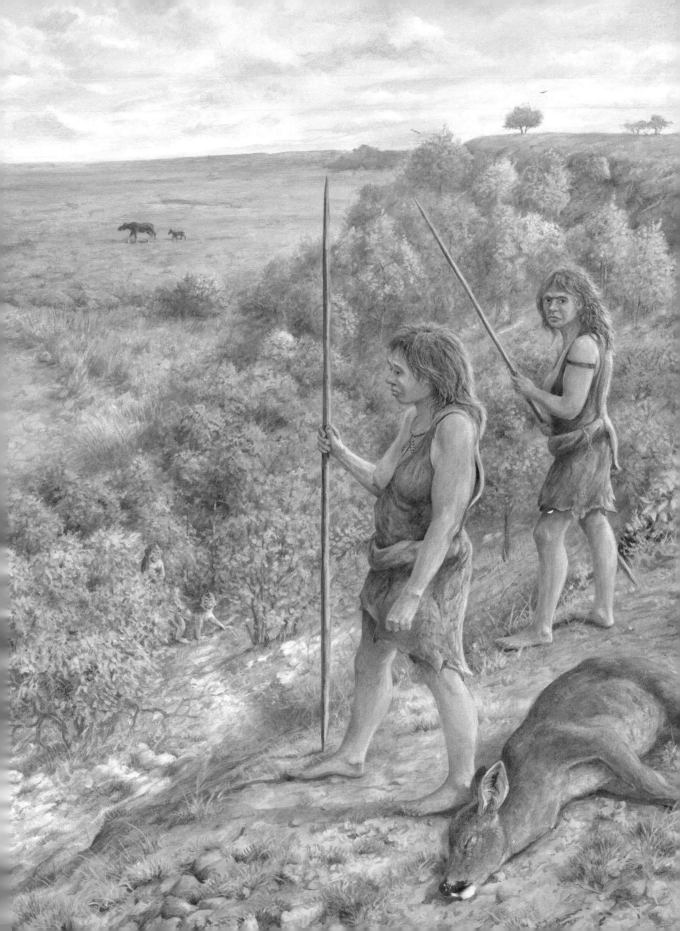

Ancient riverbank

▲ The Thames riverbank at Greenlands Quarry, an ice age hunting site at Purfleet.

◄ The chalky south bank of the ancient river.

Just out of sight, above where we stand and wonder at the contents and significance of the various gravel beds, you might imagine two hunters standing in the morning sun, keeping watch across the river and its beach. They each hold a long wooden spear. Behind them, way across the low ridge to the south, a broad grassy expanse with only a few trees enables them to spot game and watch for danger. The macaque monkeys provide an excellent alarm cry too. Below the hunters a few people work under the bankside on the gravel strand, children with an older man and a young woman. Through the silence comes the sharp, sonorous cracking and tinkle of flint being expertly worked. The woman and two of the children are quickly removing a succession of flakes from around the best pieces of flint. Above them, a scrape into the bankside has brought down a white gash of chalk with broken black flints. They are working quickly. The mother walks around to gather the worked flints in a fold of animal skin.

Suddenly, in the quiet, one of the hunters high on the bank above them makes a single, high yelping sound. They all look up and then start quickly along the shore, following two other hunters. The young lad takes the skin and runs with the flint tools up a trackway towards the top of the steep bank. He stops below the bank top and looks up at the two hunters. One of them holds a hand out a little and the boy stays still. A single call comes from way across the ridge, and one of the hunters runs down to the boy. Does she see you standing down there, in her distant future, in your hard hat and field gear, staring at her? Would

a look of utter astonishment fill her eyes? Or would she rapidly assess you as a possible meal? Maybe in our imagination and with our scientific knowledge gained from the years of excavation and research, we can look at this white riverbank now and simply wonder. We hear the roar of traffic on the M25 nearby.

▲ The Thames flowed in a loop past the Land of the Fanns and around the Chalk ridge of Purfleet 300,000 years ago. The view from the ridge has changed since then – the river has moved and modern humans have intruded with ever-increasing impact.

▼ Then and now – looking east from the chalk ridge of Purfleet towards Brentwood and Langdon Hills. Left: 300,000 years ago; right: present-day view across the same area. *Peter David Scott*

Caring for the evidence: geoconservation and Essex geological sites

Geoconservation involves identifying and caring for sites that make a special contribution to Earth heritage and which may illustrate the processes that formed the local geology and landscape. Protected geological sites can be used to generate interest and enlightenment in rock, pebbles, landscape and geological time. The general public, schools and scientific research can all benefit. Signboards, trail guides and further information for geoconservation sites help more people discover Essex rock and appreciate the scenery, land use, climate change and much more.

Geoconservation is becoming more urgent as this county's geology, rarely very visible, is becoming harder to discern. With the filling-in of old pits, ever-increasing building and urbanisation and loss of countryside, it is important to search for and define the rare sites of geological interest as valuable markers and evidence. It is just as important to ensure that such sites are appreciated and protected by county and local agencies. GeoEssex is the county's geoconservation steering group, made up of amateur and professional geologists, geographers and others concerned for the county's record of its deep history and indicators to its future. GeoEssex members seek and define sites for conservation and submit site details for protection and promotion.

Essex surface geology and its landscape are largely a legacy of the ice age. As the impacts of current climate change deepen, we can now look back over the history of this relatively short time span, with its geological record of many climatic variations, and hope to understand what is happening right now

to this planet and its life. Over historical time, for instance, farmland clearance, quarrying, building and road widening have swept countless puddingstones and sarsens stones away from the land, yet a number of these survive as standing stones, mainly on village greens, in churches and churchyards and at cross-roads. As with many other geological features in Essex, there is a constant need for vigilant geoconservation so that highway and planning authorities can be informed and updated about the importance of such evidence of geology and climate change.

Geoconservation is just one of the ways to reveal that the geological origin and evolution of the land is the ultimate basis of soil, wildlife and plants, as well as our own lives – from farming and food supply to transport, building, industry, power supply, waste disposal, mineral extraction and countless other aspects that we take for granted every day. We come from the rocks and we return to the rocks; we are part of geology and the planet's recycling.

▼ An ice age story is revealed by Gerald Lucy at the Thorndon Pebble Wall.

◄ Geoconservation in action. A wall of glacial outwash and freeze–thaw sand and gravels has been created along the edge of an old gravel pit in Thorndon Country Park south of Brentwood. Here it is being lengthened and refreshed after ten years on view to the public. Signboards have been created to reveal the surprising ice age story in the rocks.

11

An illustration from Samuel Dale's *History and Antiquities of Harwich and Dovercourt*, 1730. This reveals that a cliff of Red Crag formerly existed at Harwich. The Red Crag and its shells are at the top of the cliff and also in the fallen heap at the base. In between are layers of blue London Clay.
Reproduced by courtesy of the Essex Record Office

400 Years of Discovery

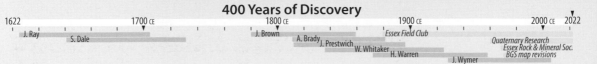

1622   1700 CE   1800 CE   1900 CE   2000 CE   2022

J. Ray

S. Dale

J. Brown

A. Brady

J. Prestwich

W. Whitaker

Essex Field Club

H. Warren

J. Wymer

Quaternary Research

Essex Rock & Mineral Soc.

BGS map revisions

Uncovering Essex geology

The Essex geology detectives

Geologists learn about the history of the Earth by studying its rocks. Both professional and amateur geologists working in and around the county have steadily increased knowledge of Essex geology and the effects of geology upon its landscape.

The rocks and fossils of Essex have aroused interest over the centuries. Rocks have been exploited to provide resources such as water and building materials and flint for toolmaking. The discovery of fossils aroused curiosity; megaliths of sarsen and pudding-stone were often set up as features in the landscape.

Early humans used the rock, especially flints, for making tools so they would have had an intimate knowledge of flint sources from Chalk outcrops and from the gravels. There are many notable Stone Age sites in Essex.

The Romans used local materials such as cement-stone for building, as well as cobbles from the gravels plus clays and brickearths for tile and brickmaking. Later these were widely recycled for use in Saxon and Norman churches. In medieval times, various materials were gathered from beaches and fields for use in building important structures such as churches, priories and abbeys.

Industrial uses for materials such as copperas (pyritized wood and pyrite nodules) from the London Clay stem from the 1600s. People discovered where to find these and a well-organised industry developed. Coprolites – phosphate nodules once thought to be fossil dung – were also located and collected, founding the fertiliser industry in north Essex. Chalk was mined to produce lime for agricultural use, while cementstone (lime-mudstone) formed the basis of the Roman cement industry. Through these activities, knowledge of economic geology, the different rock types and their occurrences, was developed. The fossils and stone artefacts found alongside these deposits caused curiosity and speculation as well as feeding the natural desire to collect things.

▼ Roman building materials – brick, cementstone and flints – recycled in the Norman church wall of St Mary Broomfield.

Interest in the study of natural history for its own sake developed in the seventeenth century following the work of John Ray, 'Father of Natural History', born in Black Notley, Essex in 1627. Although he worked mainly on the taxonomy of plants, his systematic approach greatly influenced the recording of fossils and other geological material.

John Ray FRS (1627–1705) was an English naturalist widely regarded as one of the earliest of the English parson-naturalists.
Wikimedia Commons

Copper-plate etching of shells found in the Earth from 1693 discourse by John Ray.
Essex Field Club

Samuel Dale 1659–1739.
Oil painting by unknown artist *c.*1731.
Worshipful Society of Apothecaries of London

Dale's whelk, *Leiomesus dalei.*

Samuel Dale, with whom Ray corresponded, was an avid collector of fossils from the Essex coast. Dale opened an apothecary's shop in Braintree in 1680 and practised as a family doctor in Braintree for over half a century. Dale was a staunch nonconformist and founded a religious meeting house at Bocking in 1707. The role of the gentleman and the clergyman collector stems from this period and their discoveries are still significant today. In 1703 Dale wrote an article on Harwich cliff and its fossils, based on his fieldwork which commenced in the 1690s. He refers to the Red Crag fossils at Walton. He expanded this account in his *History and Antiquities of Harwich and Dovercourt*, which appeared in 1730. He considered the origin of the fossils and the Harwich fossils and cliffs were illustrated in his book.

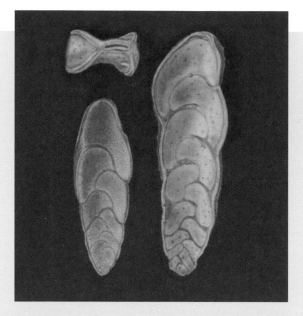

▲ The microscopic foram *Loxostomum eleyi*.
Cushman 1927. *Haydon Bailey*

A nineteenth-century vicar of Broomfield, Essex, Revd Henry Eley, became greatly interested in the new science of geology and combined this with his fascination for studying the pebbles in his garden. He invented a technique for looking at the microfossils in flint and wrote a book called *Geology in the Garden or the Fossils in Flint Pebbles* which was published in 1859. He dismissed Noah's flood, the 'Deluge of the Scriptures', saying: 'That much more recent event might leave no such trace behind it.' Instead, he extolled the writings of Charles Lyell, the pre-eminent geologist of the time, and developed the idea of glaciers shaping the land and leaving spreads of 'Boulder Drift' that had distributed the flints. He, like Dale, is honoured by having a species of fossil named after him, a planktonic foram.

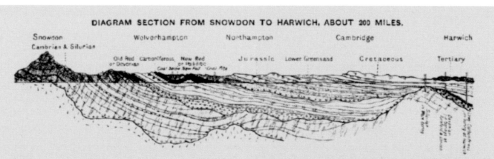

This cross section shows what would be seen in a deep cutting nearly E. and W. across England and Wales. It shows also how, in consequence of the folding of the strata and the cutting off of the uplifted parts, old rocks which should be tens of thousands of feet down are found in borings in East Anglia only 1000 feet or so below the surface.

▲ Section constructed by Sabine Baring-Gould 1910.
Cambridge County Geographies

Sabine Baring-Gould was the rector of East Mersea, Essex for ten years from 1871. He was a self-taught polymath who probably encountered the exciting new subject of geology when he was at Cambridge. His writings include geological guides and descriptions of France; detailed geological settings also feature in his works of fiction.

Much early geological interest and the collecting of fossils was (and still is today) focused on Walton-on-the-Naze. Coastal cliffs were much more accessible before promenades were built and many collections were amassed by both the learned and the fascinated. One such enthusiast was John Brown of Stanway (1780–1859). He was a stonemason who developed a great interest in collecting fossil bones of mammals from Walton Gap, the low area between Walton town and the Naze. He joined the Geological Society and corresponded with the eminent fossil collectors of his day including Sir Richard Owen, who described the first British dinosaur. Dr James Parkinson (1755–1824) visited Walton several times and records finding bones from large mammals including an Irish elk, rhinoceros and elephant. He also obtained two hippopotamus teeth that he recorded in his illustrated treatise.

John Brown of Stanway (1780–1859), a stonemason who developed a great interest in geology and collected fossils particularly from Walton.
Essex Field Club

Sir Joseph Prestwich, FRS (1812–1896), a British geologist and businessman, who was an expert on the Tertiary Period.
Wikimedia Commons/Popular Science Monthly

In the nineteenth century, the rise of industry increased pressures on raw materials such as coal and water. The search for these brought about the excavation of boreholes and the systematic mapping of geological strata, which led to the publication of detailed accounts and maps. Sir Joseph Prestwich, a City wine merchant, took an exceptional interest in the geology of London and Essex, naming the major units of rock and submitting essays to the Geological Society of London, which awarded him the prestigious Wollaston Medal in 1849. Prestwich was eventually knighted and became professor of geology at Oxford University. His large collection of British fossils, which included many from Brentwood and Margaretting, was donated to London's Natural History Museum and several examples of the molluscs from Margaretting are in the Essex Field Club's geology collection.

Prestwich and his successors, especially William Whitaker (1836–1925) of the Geological Survey, were able to take advantage of the sections revealed in the new railway cuttings through the countryside. Whitaker mapped much of southern England, including Essex, in meticulous detail. He wrote the first memoirs describing the outcrops and noted the remarkable fossils and aspects of the economic geology of Essex. He systematically curated and published a huge database of wells and boreholes that formed the backbone of hydrogeological mapping of the London Basin along with his series of water supply memoirs. He was president of several local societies and a member of many more. He was president of the Essex Field Club in 1911–14.

William Whitaker (1836–1925)
mapped much of southern England and compiled
a database of water wells and boreholes.
Essex Field Club

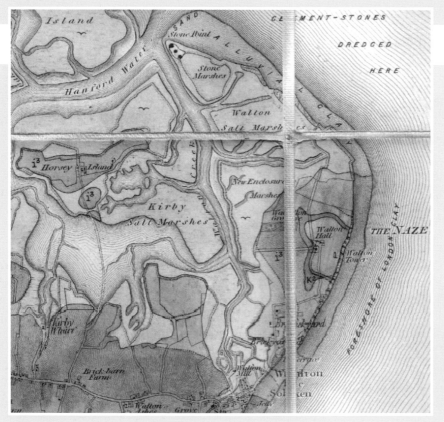

Whitaker's
hand-coloured map
of the Walton area, 1871.
Essex Field Club

T.V. Holmes (1840–1923), another gentleman geologist, took early retirement from the Geological Survey for whom he mapped in the north of England, and concentrated on describing the geology nearer to his home in Greenwich. His major contribution was the observation of Thames gravels overlying the glacial till ('boulder clay') in the new Hornchurch railway cutting in 1892.

Both Holmes and Whitaker advised on the possibility of finding workable coal deposits in Essex. Deep boreholes were sunk, first at Harwich (1854–57) in search of water, then at Stutton (1894–95), just in Suffolk, and Weeley (1896) in search of coal. None of these expensive enterprises was successful. They proved shallow, ancient basement rocks that also precluded any possibility of oil and gas reservoirs in later times.

Thomas Vincent ('Rabbity') Holmes (1840–1923) contributed many geological reports to *Essex Naturalist* including details of the Hornchurch Railway Cutting and of Deneholes.
Essex Field Club

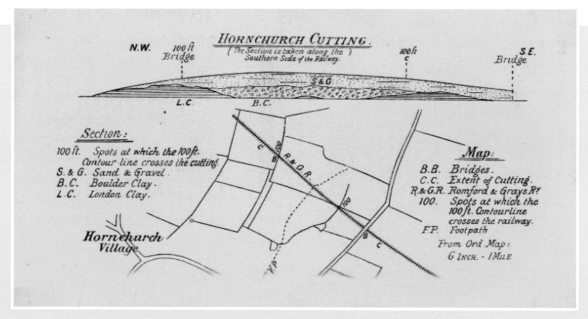

Sketch of Hornchurch railway cutting section 1893.
Essex Field Club

Another contributor to the study of geology in Essex was Samuel Hazzledine Warren (1872–1958), the son of a wholesale merchant. He entered the family business, but was able to retire in 1903 while still in his early 30s, to concentrate on geology and archaeology. Most of his fieldwork was carried out in Essex, notably at the Middle Pleistocene sites (c.400,000 years old) of the Clacton area where he identified the ancient 'Clactonian Industry' and found the tip of one of the oldest known wooden spears. He also researched late glacial deposits in the Lee Valley. He was a prominent member of the Essex Field Club, the Geologists' Association, where he received the Stopes Medal, and the Geological Society. Hazzledine Warren published many papers about his fieldwork and made significant contributions to early twentieth century debate on a chronological framework for the Quaternary (ice age) period. In 1936 he gave part of his large collection of mammal bones, other fossils and flint implements, including the Clacton spear, to the Natural History Museum in London.

Samuel Hazzledine Warren (1872–1958) found the earliest known wooden spear at Clacton and worked extensively on the Quaternary period.
Essex Field Club

The Essex Field Club, established in 1880, has many eminent fossil collectors and amateur geologists among its past and present membership. The Essex Field Club also houses the principal collection of Essex rocks, fossils, books and geological site records. Similarly, the Geologists' Association, from its inception in 1858 to the present day, has held many field excursions to view geological phenomena in Essex. Accounts of recent researches into Essex geology still feature in the pages of its Proceedings.

Blue plaque to Hazzledine Warren who lived at 49 Forest View Road, Loughton, Essex.
Gerald Lucy

A Geologists' Association visit to Brickyard Pit near ► Wenden Mill, Audley End station, 22 July 1911.
Geologists' Association Carreck Archive,
Reader Geological Photographs,
British Geological Survey © UKRI 2021

John Wymer (1928–2006) was an enthusiastic and engaging discoverer of human artefacts in Essex, described as Britain's foremost specialist in Palaeolithic archaeology, with an unparalleled knowledge of the Quaternary. In 1954 he was excavating at a Palaeolithic site at Little Thurrock to investigate the puzzle of the absence of hand axes at Clactonian sites. He published widely and was able to put the many Essex sites he investigated into a wider context.

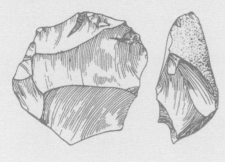

John Wymer (1928–2006),
Britain's foremost specialist in Palaeolithic archaeology,
with an unparalleled knowledge of Quaternary
geology and the earliest human artefacts.
Essex Field Club

Fossils discoveries from brick pits

The lack of suitable building materials in Essex led to the manufacture of bricks and tiles from a very early date, using local clays and brickearths. The Romans made bricks extensively for their buildings which, in turn, were subsequently used as 'quarries' for building material during later periods. As the towns, and particularly London, expanded not only did every settlement have its own brick works, but many tons of bricks were shipped to the City and the East End. Various geological formations were used and the coincidence in Essex of clay with suitable sand was most fortunate. In south Essex many thousands of workers were employed making bricks from the local brickearth deposits for building Martello towers as well as for housing. As these were dug by hand, many spectacular fossils were discovered.

▲ 'Jubilee Trucks' being loaded with clay, Salvation Army Brickfield, Hadleigh, Essex, c.1915.
Peter Gillard, Benfleet Community Archive

The discovery ▶ of mammoth and other fossils at Ilford by Antonio Brady is commemorated by a plaque on the wall of Ilford Methodist Church in Ilford Lane.
Gerald Lucy

MAMMALIAN REMAINS.
SITE OF UPHALL BRICK-PITS
FROM WHICH NUMEROUS PLEISTOCENE MAMMALIAN
REMAINS WERE UNEARTHED BY
SIR ANTONIO BRADY. c.1850.
THE COLLECTION IS NOW IN THE POSSESSION OF
THE NATURAL HISTORY MUSEUM.
THIS TABLET WAS PLACED HERE BY
THE ILFORD BOROUGH COUNCIL, ON BEHALF OF
THE CITIZENS OF ILFORD, TO COMMEMORATE
THE FESTIVAL OF BRITAIN. 1951.

The wife of the owner of an Ilford brick pit would send a letter to a local collector and enthusiast, Antonio Brady, when the quarry workers uncovered any bones. His finest specimen, found in 1864, was a complete mammoth skull with both tusks. He spared no effort to excavate the fragile fossil and donated the specimen to the Natural History Museum, where this and many other of his specimens remain today.

▼ The wonder of the Ilford mammoth. *Gerald Lucy*

▲ Sir Antonio Brady (1811–1881), collector of mammal fossils from the Pleistocene brickearth and gravels of Ilford, Essex. *Essex Field Club*

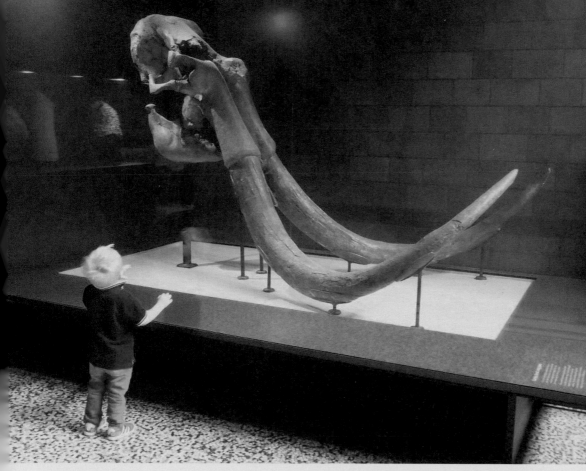

Modern geological detective work

In the 1970s, geological mapping revision was carried out by Bristow, Ellison and Lake, among others, of the Geological Survey. This involved the drilling of new boreholes to investigate layers down to the Chalk in mid-Essex. Examination of the borehole cores led to a revised description of the Paleogene strata. Further detailed studies by Dr Chris King resulted in a large report, *A Revised Correlation of Tertiary Rocks in the British Isles and Adjacent Areas of NW Europe*, published by the Geological Society in 2016. Revisions to nomenclature by the BGS followed. Name changes for rock layers can be confusing. However, with more research comes a greater understanding of how each 'time-layer' is correlated and changes in the names of various beds need to be made from time to time.

In the 1970s and 1980s excavations for major motorways across Essex, the M11 and the M25, created further opportunities for research. As well as many interesting fossils, complete sections through the geological strata were logged and recorded to add to previous knowledge. Graham Ward and many amateurs made collections, some of which are lodged with the Essex Field Club. Considerable research from the 1970s onwards also focused on the more recent gravel and glacial till deposits across Essex, both to assess the gravels as a commercial aggregate resource and to study them scientifically in order to clarify the ice age story of Essex. Following the lead of Dr Richard Hey in the 1960s, research into putting archaeological studies of the Palaeolithic (Old Stone

▲ Dr Peter Allen at Bulls Lodge Quarry.
He carried out pioneering research into Thames terrace and glacial outwash gravels and till deposits in north Essex and Suffolk.
He leads many field visits to the gravels of Essex and helps to enlighten others as to their interpretation.

▲ Professor Phil Gibbard (left) established the detail and timing of the diversion of the Thames and was instrumental in defining the Quaternary period.
On the right is the late Professor Richard West who analysed pollen to distinguish and name Pleistocene interglacial stages.
P. Gibbard

Age) into their geological context led to an approach based on the kinds of pebble in each gravel layer. This heralded a new era in the study of the history of the River Thames. Professor Jim Rose pioneered the use of palaeosols (ancient soils) in British Quaternary (ice age) geology. He also recognised the scale of river activity in relation to climate change. Further university-based research enabled geologists to distinguish river gravels from gravels that had been washed out of the ice sheet across the region; previously these had all been described and mapped as one unit. Dr Peter Allen, working mainly in Suffolk, identified and named the Gipping Till and recognised warm- and cold-climate soils in the underlying gravels. This work was extended into mid-Essex by Dr Colin Whiteman.

Following ideas early in the twentieth century that the Thames had flowed north-eastwards though the Vale of St Albans, Phil Gibbard established from the 1970s, the details of timing and sequence for the diversion of the Thames into a southern Essex course. The progressive quarrying of Thames sand and gravel for aggregates exposes ever-changing sections of this ice age geology, crucial for discerning the past and indicating the future. Regular access to quarry faces by geologists is highly valuable to the future of ice age research.

▲ Dr Richard Hey (1917–2011) on 400 ft Pebble Grave, Northaw Great Wood, Herts, 1977. He pioneered the characterisation of gravels using the composition of the pebbles.
P. Gibbard

▲ Professor Jim Rose at the Thorndon Pebble Wall. He recognised the use of palaeosols to aid the correlation and dating in relation to Quaternary climate cycles.

Confirmation that the River Medway was an integral part of the Thames river system was pioneered by Professor David Bridgland; he subsequently developed a widely used template for river terrace formation from work in south Essex. He was also the first to tie the terrace sequences to the marine isotope stage ocean (MIS) record. Professor Danielle Schreve was then able to identify and construct mammal assemblage zones (MAZs) to correlate interglacial deposits and to identify MIS sub-stages of temperature variation.

Investigations along the route of the High Speed 1 rail link (HS1) through Purfleet were particularly significant. These studies established south Essex as having the best post-Anglian ice age record in Europe, if not the world. In 1997 Professor Schreve identified bones found at Aveley as those of the first jungle cat discovered in Britain. This was in addition to the two mammoth/elephant skeletons found nearby by amateur fossil collector John Hesketh in 1964.

Gerald Lucy, through diligent work over a long period, has reviewed all the geological literature relating to Essex and visited most of the locations cited. He has produced a comprehensive gazetteer of geological sites in Essex, available through the Essex Field Club website. This has provided GeoEssex, the county's geoconservation steering group, with the basis and means to identify geoconservation sites and to notify county planning authorities of their significance.

▲ 'All for the love of gravel' – Professor David Bridgland in ground-breaking research, using gravel pebble composition to distinguish and correlate river terraces. *David Bridgland*

▲ Professor Danielle Schreve established MAZs to correlate interglacial deposits. *Danielle Schreve*

▲ As a life-long enthusiast for Essex geology, Gerald Lucy continues to compile historical, landscape and geoconservation evidence.

Continuing discovery

Discovery and identification of new fossil species still happens. Amateur collector Bob Williams lodged a fossil find from 1991 with the Natural History Museum. Later, new research techniques at the Museum identified it as a new species. This Eocene puffer fish, *Ctenoplectus williamsi*, was named after him in 2016. Other amateur geologists have made significant contributions to Essex geology in recent times. Fossil collecting is still a popular hobby, though many of the past favourite localities have now been land-filled, eroded or otherwise made inaccessible. An important collection of fossil bird bones from the London Clay at Walton-on-the-Naze has been made by Mike Daniels. Bill George has identified many fossil localities, especially along the coast; he has made significant finds of mammal bones, fish fossils and flint artefacts as well as chronicling the history of the study of geology in various localities. Jeff Saward and others have made extensive collections of London Clay crustacea and shark teeth, through which much has been learned about the palaeoecology of the time.

New tunnelling works for major engineering projects such as Crossrail and the Thames Tideway scheme have enhanced the understanding of variations within the geological strata at depth. This has shed more light on structural features such as folding and faulting in the south and west of the county.

Boreholes drilled in the nineteenth century showed that the hoped-for economic targets such as coal and oil or gas reservoirs do not exist under Essex. Hence the nature of the ancient basement rocks beneath the Chalk remains largely unknown. Much has been learned from the exploitation and study of the rocks of Essex, but there is still much more to investigate and understand.

▲ *Ctenoplectus williamsi*, found at Aveley in 1991 in a London Clay nodule and named in 2016. *Bob Williams*

A tunnel boring machine ►
for the Thames Tideway project. *Tim Newman*

12

Grading sand and gravel in a modern quarry south of Stanway near Colchester. Used as aggregate for the construction industry, this is taken from glacial outwash and underlying Thames gravels. Summer floods from the Anglian ice sheet spread a delta of sand and gravel across the whole of the Stanway district. Deposited beyond the ice front, this layer has no overburden of glacial till, making it easier and less expensive to extract.

Using the Land

1,000 years | | | | | | 500 | | | | ► Water from Chalk wells | Now
► Cobble walls *recycled stone & Roman brick* | | | | ► Limekilns | ► Brick towers | | ► Copperas | ► Harwich cementstone | ► Aggregate
| | | | | ► Flint flushwork | | | ► Phosphate | Industry

Rock and people

Economic geology: what we take out of the ground

All across Essex, there are hundreds of names of roads and lanes, farms and fields, localities and housing developments that tell of past industries. They reflect the social and commercial importance of local geology. Limefields, Kiln Lane, Pottery Lane, Tile Kiln Estate, Sandpit Lane, Brickfields, Marlpit Road and Copperas Road all tell of industries both ancient and recent.

The whole range of Essex rock, described throughout this book, has been used to feed the growth of civilisation in and around the county throughout the last few thousand years, from the Chalk Formation up to the present-day alluvium. Everything we use that cannot be obtained from the sea and air or from plants has to be taken out of the ground. Uses of Essex geological materials range from flint for toolmaking, which has been used through almost the whole of human prehistory, through to fashioning clay for pots and bricks, quarrying stone and gravel for walls and roads and digging chalk to make lime. Today, Essex is the largest producer and consumer of sand and gravel in the east of England. There are 20 permitted sand and gravel sites, one silica sand site, two brick clay and one chalk site.

Clay and brickearths have long been used to make tiles and bricks and, in more recent times, clay, chalk, sand and gravel were together used to make cement and concrete in ever-increasing quantities. Large quarries existed close to the border of Essex and Suffolk where both the chalk bedrock and the overlying clay and gravel layers are accessible. These fed the growth of industry, urbanisation, farming and road building. Through medieval times and into the years within living memory, there have been very large numbers of

◄ A worked flint tool and a naturally broken flint.
John Ratford

▲ Handmade bricks, made from London Clay,
at Bulmer Brick & Tile works.

Peter Minter passes on the Brickmaker's Tale. ►

pits and quarries of all sizes across the whole of Essex.
Enormous exports of bricks and cement continued
until production elsewhere supplanted most of the
Essex industry through the twentieth century. The
uses and knowledge of geology in engineering and
construction, for tunnels, bridges, embankments,
excavations, road building and foundation engineering
are now of ever-increasing importance as significant
additions to the county's infrastructure, as well as more
home construction, are planned.

Flint and stone

Across southern England and into Essex and the North Sea area, countless billions of flints have been released as rising uplands of soft chalk were very

▼ A seam of fresh flint in chalk at Chafford.

rapidly eroded away. They were rounded into pebbles on beaches and redistributed by the Thames and other rivers to be left behind as terraces of gravel that track the ancient river paths across the land. These flint-rich gravels have profoundly affected the lives of people ever since humans first travelled here: they provided the tools for hunting and carving throughout practically the whole of human existence in the area; they provided building material for walls and roads from Roman times onwards; and now they provide resources for one of the largest industries in Essex, that of aggregate for concrete production, building and road construction.

The making of tools from flint by Stone Age people must be the most ancient extractive industry in Essex. Large numbers of such tools have been found throughout the county. Tools were made from good-quality flint extracted directly from the Chalk in those few areas of Essex where this was exposed, such as along the ancient Thames riverbanks near Purfleet in south Essex. Elsewhere across the land, people would have found sites for good quality flint in gravels.

Flint was later used for the manufacture of gun flints. This was carried out at Purfleet until the middle of the nineteenth century when flintlock weapons were finally superseded by those with percussion locks. Another industry was the calcining of flint, which formerly took place at West Thurrock to supply the potteries at Stoke-on-Trent. The good-quality black flint was heated and finely ground for use as a filler in whiteware ceramics.

◄ Flints in shattered Chalk,
    Greenlands Quarry, Purfleet.

Early sixteenth-century flint and limestone ►
flushwork, Chelmsford Cathedral.

Building stone

Good stone for building is very scarce in Essex and many early buildings had to be constructed with whatever rocks were available locally. Cobbles from the fields were bonded together with lime mortar made from chalk and sand, sometimes together with broken flints and seashells. The commonest natural building material for walls, cobbles of flint with other materials such as quartzite and vein quartz, were cleared from the land and dug from local pits. Much of this material was deposited by the ancient River Thames and its tributaries that once flowed across north and mid-Essex. The walls of many churches may be described as 'rubble walls' or even 'unremarkable flint work', yet closer examination reveals the fascinating variety of materials that make up the cobbles.

The construction of round church towers – there are still six in Essex – came about because of the difficulty of making satisfactory corners (quoins) with locally available materials. Material for medieval churches was often 'quarried' from nearby Roman buildings including bricks and tiles and cementstone as well as cobbles of Thames gravel. Each wall is different. Church and tower walls, garden and cottage walls and estate walls all provide clues to the fascinating story of Essex geology. Indeed, they are the most easily observed rock outcrops across the county.

▲ Broomfield St Mary
Recycled Roman brick and cementstone
– pale ginger coloured.

▲ Faulkbourne St Germanus
A block of tufa from a local stream-bed
deposit. Similar to travertine.

For wealthy churches and other important buildings throughout Essex, 'knapped' flints were used as facing stone – flints skilfully worked to produce a flat, black surface. This was revived in churches built in Victorian times. Knapped flint, together with chalk or limestone, was also used to make decorative flushwork. Fine examples of this craftsmanship can be seen on the fifteenth century gatehouses of St John's Abbey Colchester, St Osyth's Priory, near Clacton and in Chelmsford Cathedral. The best quality flint for knapping, however, probably did not come from Essex, but from Brandon in Suffolk, an area where flint has been mined since the Stone Age.

Knapped flint wall, Dedham Church. ►

▲ Little Bardfield St Katherine
Saxon tower with local sarsen, puddingstone and large flints in quoins, plus big sarsens in the tower face.

▲ Little Baddow St Mary
A large proportion of smooth quartzite cobbles, with a puddingstone and recycled stones.

Limestone from elsewhere came into use when better transport and more money enabled supplies of shaped building stone and monument stone to be brought into the area. Meanwhile, clay was fired to make brick and tile. Finance dictated the use of stone, brick or recycled cobble material in churches or other important building work. The closest limestone sources were Kentish Rag (Lower Greensand) from the Maidstone area, Reigate Stone, Lincolnshire Limestone from various locations, including Barnack in early medieval times, and Chalk Rock from an area just to the north-west of Essex; Caen Stone from Normandy was also used in Essex. Most old walls contain a large amount and variety of recycled building material, including previously used limestone blocks, cement-stone, ferricrete, sarsens and Roman bricks.

Cementstone, also known in the form of septarian nodules or septaria, is found in the London Clay. This material can be seen in old buildings across east to central Essex, particularly in churches close to the coast. The Romans used cementstone extensively, a surviving example being Colchester's Roman wall, but probably the best example of the use of this stone is the Norman castle at Colchester. Large amounts of cementstone were quarried from cliffs, shores and offshore at Harwich and nearby.

In north and west Essex there are churches constructed with Hertfordshire puddingstone, sarsens and various erratics. Further to the south and east there are churches built using blocks of ferricrete, an iron-cemented sand or gravel, which makes a remarkably long-lasting wall stone. An excellent example of the latter is in the church at Great Bentley, near Colchester.

◄ Dark brown ferricrete in the wall of St Mary, Great Bentley. *Gerald Lucy*

As a building material, Chalk seems an unlikely choice, but at some horizons in the formation a harder chalk known as 'clunch' occurs. This was quarried in the extreme north-west of the county. It was fairly soft when quarried and sometimes left for up to two years to dry out and harden sufficiently before use. Large amounts of this material have been used in the interior of Saffron Walden church and a magnificent carved clunch mantelpiece can be seen in Saffron Walden Museum.

Chalk and lime

Wherever the chalk occurs at or near the surface, it has been dug in pits and quarries of various sizes to provide material to improve the soil and for many building purposes. Applying chalk to the soil to improve fertility has been practised since Roman times – evidence associated with Roman archaeological sites has been found in north Essex. In 1225 Henry III gave every man the right to sink a marl pit on his own land. Many farms on the boulder clay plateau of north Essex originally had at least one 'marl pit' for improving the soil, especially when pasture was being converted to arable land. Spreading chalk on the fields was a common practice in the Middle Ages. The heavy London Clay soils of mid-Essex were chalked to reduce acidity and make the clay more workable. Chalk was also used to improve the fertility of light sandy soils developed on the Thanet Sand in north and south Essex and on the lower Thames gravel terraces. Agricultural lime, or crushed chalk, is the main product from the Newport Limeworks in north-west Essex, the only remaining active lime quarry in the county. The quarries at Purfleet were originally dug for agricultural lime before being taken over for cement production. Chalk was also used in paints, medicinal tablets and the foundations of buildings and farmyards. Rammed chalk made a very hard base and is still used as a base for cattle yards.

Deneholes

Although there is speculation as to their original purpose, it has been suggested that the deneholes at Hangman's Wood near Grays and elsewhere in Thurrock were dug through the Thanet Sand into the Chalk for the extraction of chalk in medieval times. They consist of vertical shafts through the Thanet Sand and end in branching chambers cut into the underlying chalk. The Hangman's Wood deneholes are particularly deep; the shafts are over 20 metres deep before the Chalk is reached. The Essex Field Club carried out the first extensive investigation into the nature and origin of these deneholes and published a comprehensive report in 1887. At that time 51 shafts were known at Hangman's Wood but all except five were blocked and could only be identified by depressions on the surface. The Field Club carried out further investigation of the site in the 1950s and early 1960s. It is thought that there may be as many as 72 shafts, or deneholes, on this site. Some deneholes were subsequently destroyed by chalk quarrying. Others have been the cause of sudden collapse in recent times. One was recorded in Blackshots Lane, Grays in 2012 when a 7 m (25 ft) wide hole suddenly appeared in the road.

▲ Descending a denehole at Hangman's Wood, Little Thurrock. 'In the course of ploughing this field the horse's feet broke through the surface. We procured a rope and the pole of a waggon and investigated the interior. The shaft descended through about 12 feet of Thanet Sand and entered into a beehive-shaped chamber about 20 ft. high. These deneholes were formed in getting the chalk in past ages for marling the land.'
Geologists' Association Carreck Archive,
Reader Geological Photographs.
British Geological Survey © UKRI 2021

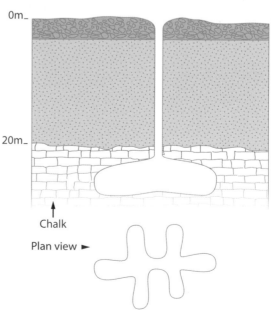

0m‾

20m‾

↑
Chalk

Plan view ►

Section and plan of a typical ►
denehole in south Essex.

Limekilns

Chalk was 'burnt' in kilns for the production of mortar and many other uses including in tanning, lime-washing walls, cleaning wooden utensils, preserving food such as hams, disinfecting hen houses and disposing of animals that had died of disease. The earliest record of a kiln in north Essex is from the time of Henry VI in 1425 when tithes from a Bulmer limekiln were apportioned to a newly appointed vicar, but lime burning was practised in Roman times. In 2016 a limekiln thought to have been used to make the lime mortar during the construction of Henry VIII's Beaulieu Palace, now New Hall just north of Chelmsford, was discovered prior to housing development.

There were, no doubt, numerous limekilns throughout Essex, some close to chalk pits but mainly at sites where the product was required. The only complete limekiln that survives in Essex is at Beaumont Quay, near Thorpe-le-Soken. Lime was mainly used for the manufacture of lime mortar, used for construction with cobble and brickwork, as well as for render. It is flexible, age-hardens and allows water vapour to pass through, enabling a building to 'breathe' unlike, for instance, modern Portland cement.

▼ Maldon limekiln at the start of the twentieth century. A painting by Mary Elizabeth Stormont. *Rye Art Gallery Permanent Collections*

Salt

The extraction of salt is a very ancient industry. The Romans are known to have recovered salt at many sites on the Essex coast by evaporating sea water in large shallow dishes of baked clay. The visible signs of this industry are the 'red hills' that can be seen along shores and estuaries; these were formed from an accumulation of red soil which was a by-product of the fires used to apply heat to the salt pans. Salt is still produced at Maldon by the Maldon Crystal Salt Company.

▲ Red soil beds from salt making at the Naze. *Bill George*

Copperas

The little-known copperas industry was formerly of great importance in Essex. It consisted of the gathering of pyrite nodules and pyritised twigs, known as 'copperas stones', from Essex beaches, where they had been washed out of the London Clay. These were allowed to oxidise for several months in open vats, whereupon a solution of ferrous sulphate, 'green vitriol', was formed. This was an essential chemical for dyeing leather and cloth black and for making black ink. Women and children were employed to gather copperas from the beaches and were paid in 'copperas tokens'. Old records show that over 230 tons a year were removed from the beach at Walton in the period 1715–1720. Daniel Defoe, in his *A Tour thro' the Whole Island of Great Britain* (1724), noted: 'At Walton, under the Nase, they find on the shoar, copperas-stone in great quantities; and there are

▲ A Walton copperas token, issued by John Kidby, rector of Shenfield about 1680, later counterstamped 1736. *Stuart Adams*

several large works call'd Copperas Houses, where they make it with great expense.'

Although Walton was a major centre for the industry, it was also carried out at Frinton, Harwich, Wrabness and Brightlingsea. It died out over 100 years ago, but evidence of it still survives, for example 'Copperas Bay' and 'Copperas Wood' at Wrabness.

▲ Pyrite-rich nodules and fragments of fossil wood from the London Clay were gathered in large quantities as 'copperas stones'. These were converted to 'vitriol' for a variety of uses.
Billie Bond

▲ Road name in Brightlingsea.
Gerald Lucy

Phosphate 'coprolites'

The fossilised droppings or excreta of prehistoric animals are called 'coprolites' and these are occasionally found in some sedimentary rocks. The phosphatic nodules within the nodule bed at the base of the Red Crag were originally thought to be coprolites and it was formerly given the name 'coprolite bed'. These are in fact phosphate nodules that formed on and in seabed sediments from fish and other animal remains. In Suffolk, these nodules were the basis of an agricultural phosphate fertiliser industry dominated, in the middle of the nineteenth century, by firms such as Fisons. When the benefits of these coprolites were directly identified, there was a 'Coprolite Mining Land Rush' according to the Museum of Rural Life at Stowmarket, although Red Crag had been spread on the land for many years before in an attempt to improve its fertility. In Essex, these nodules were worked on a smaller scale at Wrabness and Walton. Phosphate fertiliser production was an important British industry before plentiful overseas discoveries were made.

◄ Phosphate nodules, known as coprolites, were dug out and gathered to crush for use as valuable fertiliser.

Clay lump

'Clay lump' is the colloquial name for unfired clay bricks which were used to make walls for cottages and farm buildings. Provided they are on brick or stone foundations, once plastered and whitewashed, the buildings were warm and durable. Indeed a few still survive in north Essex. The clay was mixed with a little sand and straw, then 'trodden in by a horse' before being thrown into wooden moulds, then dried by wind and sun. Moreover, they were not liable to brick duty, though this was repealed in 1850.

Bricks, tiles, drains and pots

Pottery has been made from the geological layers of Essex, notably its clays, with the addition of crushed flint from gravels, since early Neolithic times around 6,000 years ago. Brick and tile making was introduced to Britain by the Romans nearly 2,000 years ago and their bricks, because of their durability and shape, have been reused in the structures of many ancient Essex churches. The dimensions of Roman bricks vary but they can be recognised by their characteristic shape, being rather more tile-shaped than present day bricks. Following the departure of the Romans, brickmaking was not carried out again in any great quantity in Britain until the twelfth century; however, some of the earliest surviving examples of post-Roman brickwork can be found in Essex, and Saxon bricks have been recognised, with their distinctive size and slightly bowed shape due to drying while laid out on the ground.

Bricks and tiles, plus pottery, were once made all over Essex at many hundreds of works. In an area devoid of natural building stones, bricks were

▼ Brickwork at the height of its status – Layer Marney Tower, England's tallest Tudor gatehouse built in the reign of Henry VIII by Henry, Lord Marney. Repairs have been made with hand-made matching bricks from Bulmer Brick & Tile works in north Essex. *Nicholas Charrington*

▲ Benjamin Perkins's depiction of the building of Gestingthorpe Church tower, probably in the mid-1520s. Mortar is being mixed for laying the new load of bricks made close by. The chalk, sand and clay for these were all dug from the land locally, as were the flints for the church walls. *Ashley Cooper*

manufactured from a variety of local deposits. Relatively few Essex parishes have no evidence of brick, tile or pottery manufacture. In the nineteenth century, virtually every town had its own brickyard and there were still a hundred brickyards in Essex before the Second World War. Pits were dug into the London Clay and in places into the Reading Formation below, the Claygate Beds, glacial till ('boulder clay'), 'head' and the various brickearths, alluvium from river valleys and sand beds across the county. Chalk was also mixed with the clay in processes developed in the nineteenth century. Large brickmaking industries supplied bricks, flooring and tiles, and drainage pipes far and wide, with fuel supplies and exported products coming and going by river and, later, by railway.

Brickmaking developed rapidly from Tudor times, perhaps using improved methods, and red bricks became fashionable. Brick-built towers were added to churches and towers were rebuilt with bricks after they fell into disrepair. More churches in Essex have brick towers than in any other English county. In the late seventeenth century Daniel Defoe, author of Robinson Crusoe, owned and operated a successful brickworks at Tilbury. In 1804 about 500 men were employed digging the Grays brickearth to make bricks for Martello towers.

Increasing numbers of bricks were made as London expanded and for the building of the railways. At Ilford, the extensive brickearths with seams reportedly 6 m (20 ft) thick were dug by hand and the bones of many mammals were found, the first recorded in 1786 at Uphall. These included the Ilford mammoth found in 1863 in the area that became known as the 'Ilford Elephant Ground' with the discovery of the remains of more than 300 elephants.

Farmers and estate owners would open a brickyard for one specific project. All 6 million bricks needed for the viaduct at Chappel were made at brickworks within a mile of the site.

The Bulmer Brick & Tile Company is the sole remaining working brickworks from the very large Essex-Suffolk borders industry. It is a small family

London Clay being ► dug by hand at Bulmer brickworks, about 1910.
Photo provided by Adrian Corder-Birch

business continuing the traditions of brickmaking on this site dating back to the Middle Ages. They survived the Second World War by making land drains and bricks for the many local airfields. Today they make specialised hand-made products for restoration and rebuilding projects, both small and large, such as St Pancras station and Hampton Court Palace. They use the sandy clay near the base of the London Clay – the Wrabness Member of the Harwich Formation – together with some clay from the underlying Reading Formation. They are able to make mixes of various clays and match existing bricks for historical restoration purposes, such as at Layer Marney Tower.

At Marks Tey, near Colchester, a grey clay from the ice age lake bed sediments is worked by W.H. Collier Ltd in a small brickworks that was established in 1863.

▲ The London Clay face being dug at Bulmer brickworks in 2014. The clay is dug before autumn, ready for weathering for some months to remove pyrite and improve the texture.
Peter Minter, Bulmer Brick & Tile Co. Ltd

A pair of downdraft kilns still survive from 1883. The clay pit is important geologically and is protected as a Site of Special Scientific Interest.

Brickearth based on wind-blown loess was worked for brickmaking until quite recently at Great Wakering, where the Star Lane brickworks was dismantled only in 2007. This deposit was once extensively used for brickmaking in the Southend area. The orange loess in this area is of particular scientific interest because it is 'primary loess'; in other words, it is an original wind-blown (aeolian) deposit that has not been reworked by water.

Essex Rock and ► Mineral Society 50th Anniversary brick made from London Clay by Bulmer Brick & Tile Co. Ltd.

At Hadleigh in south Essex there was a major clay extraction and brickmaking operation from the late nineteenth century up to the 1960s. Set up by the Salvation Army in 1891 as part of a pioneering social programme to empower people from the East End of London who were living in poverty and experiencing long-term unemployment, brickmaking, together with farming, sustained the colony. The land provided the raw materials of brickearth from the Claygate Beds, London Clay and Bagshot Sand. The pits are now flooded and used for recreation. In 1898 around 50,000 bricks a day, of many types, were produced during the summer months. A private railway line provided a link to the estuary over the existing Tilbury to Southend Railway line (remnants of the bridge can still be seen from the train). Barges would carry the bricks to London and return with cargoes of chalk (which was used in the making of yellow bricks) as well as rubbish from London (some of which was burned to fire the kilns). Many older people living in the local area today remember the brick fields and the people that worked there, and the distinctive bricks, impressed with WB or SALIC, are seen in local walls.

Large scale industrial brickworks based on the Jurassic Clays of Bedfordshire and around Peter-borough displaced brickmaking in Essex after the Second World War.

◄ Recycled local bricks and brick clinker in a Hadleigh garden wall. The juxtaposition of suitable transport and ideal geology – the sequence from the London Clay up to the Claygate and Bagshot beds – in the Hadleigh area west of Southend was exploited in a large industrial complex during the nineteenth and twentieth centuries, providing millions of bricks for building in London.

Cement

Before the invention of Portland cement in 1850, 'Roman cement' was made in Essex from accumulations of cementstone or septarian nodules washed out of the London Clay cliffs along the coast. Harwich was the main centre of this industry, where the 'cement stones' were excavated from the cliffs; later, when the supply of stone dwindled, dredging offshore was carried out to collect the stone that had accumulated on the seabed. During the early part of the nineteenth century up to 500 men were employed in this industry at Harwich and the cement was supplied to all parts of Britain and northern Europe. Much of the external rendering known as stucco used during the Regency Period was made from this cement. There was a cement factory at Leigh-on-Sea and the collection of cement stones here, and at Harwich, caused great concern because of the effect it was having on the erosion of the cliffs.

For many residents of Grays, their memory of the town is buildings covered in grey dust from the local cement industry. The cement industry was an important part of the local economy from the middle

▼ The giant Tunnel Cement Works at West Thurrock in 1963. The working chalk face can be seen in the distance and the Dartford Tunnel approach road is in the foreground. The works closed in 1976 and the site is now occupied by the Lakeside Shopping Centre. *RTZ Estates Ltd*

of the nineteenth century following the invention of Portland cement. The raw materials of chalk, clay and water, together with the railway and transport links to the Thames made for an expanding industry. In 1872, naturalist Alfred Russel Wallace arrived in Grays and built his home, the Dell, using concrete – one of the earliest such in the country, now a listed building. The rotary kiln changed the scale of cement production, previously done by hand. Tall chimneys increased the draught and waste particles were a great nuisance, covering everywhere with a blanket of grey dust.

Huge quantities of chalk were excavated over the years – 5 billion cubic metres – with nearly 100 deep holes covering 1,600 hectares (4,000 acres; 9% of Thurrock). During the 1970s other countries produced cement, lowering prices, and the supply of chalk from these quarries became exhausted, leading to the closure of factories and loss of jobs. Thurrock was scarred with a landscape of quarries, railway tracks and abandoned buildings.

Clay, the other essential ingredient of Portland cement, was worked in large pits in the London Clay at South Ockendon and Aveley. Clay was mixed with water and turned into a slurry so that it could be piped to the works. When cement manufacture ceased in Thurrock, the slurry was pumped under the Thames to works at Northfleet in Kent. All the pits have now been landfilled or converted to leisure facilities such as fishing lakes and outdoor activity centres. When they were working, a total of almost 50 m of lower and middle London Clay was exposed. Thousands of fossils such as molluscs, bird bones, fish remains, echinoid spines and turtle bone fragments, shark and ray teeth, fossilised seeds and fruits have been found there.

▼ An early twentieth century concrete wall in North Stifford made with pebble gravel from the Lower London Tertiaries quarried nearby. This is the same type of black flint pebbles found in the puddingstone and ferricrete from the same location and seen in the parish church nearby.

Aggregate: sand and gravel

Large-scale quarrying in Essex has now switched to extracting sand and gravel for concrete, building and roads – known as 'aggregate' in the industry. This is of ever-greater importance as Essex is one of the largest producers in the UK and the majority of the aggregate is used within the county itself. Sand and gravel quarries are situated all across the county, such as around Great Dunmow, Chelmsford and Colchester and in south Essex at Orsett and Fairlop, now in the London Borough of Redbridge. The gravels worked are mostly from the ice age Kesgrave Sands and Gravels, glacial outwash gravels and later river terrace deposits. The gravels consist largely of flint pebbles originally derived from the Chalk, but they also contain a proportion of rocks derived from distant sources. Some sands and gravels lie beneath a thickness of glacial till 'overburden', which has first to be removed before extraction of the aggregate can proceed.

Old gravel pits have been used for landfill, fishing lakes, golf courses, housing estates and nature reserves, or returned to agriculture, often in accordance with pre-war licences. Increasingly, companies are required to make provision for extensive and careful restoration following extraction. Seabed gravels, often tidally reworked from ancient river terraces that extend from the mainland, are an increasingly important source of aggregates for the construction industry.

A number of Essex aggregates quarries lie across former airfields such as Boreham (Bull's Lodge), Bradwell near Braintree, Highwood near Dunmow, Fairlop, and many more. The airfields were there because the gravels made suitably flat heathlands, and the till plateau was flat enough and suitable after extensive drainage.

◄ Aggregate extraction – sand and gravel quarried, graded and prepared for use in concrete, building and roads – is one of the largest industries in Essex.

In 1813 Stanway's heathland of dry, sandy and gravelly loam was ideal for turnips, but the yields of wheat, barley and oats were all below average for the county. The naturally infertile sandy gravel needed heavy manuring with Colchester town muck. The lord of Stanway manor had a gravel pit here in 1695, and in 1817 there was a larger pit on the former heath. By 1977 planning permission had been given for gravel extraction from over 120 hectares (300 acres) at Warren Lane and Stanway Hall. Church Lane quarry was one site of extensive extraction over many years and the large Tollgate Centre Shopping Park was built on part of the former Judds Farm gravel pit south of London Road between 1988 and 1990, and expanded in 1995. Part of the remaining Thames sand and gravel can still be seen under the edge of Church Lane.

The Wivenhoe Sand, Stone and Gravel Company was formed in 1925 when the owner of the agricultural land at Wivenhoe realised that the sand and gravel under his land could be exploited. Prior to that small ballast barges used the stream to load at Ballast Quay. Early working was entirely by hand with horses to carry cartloads of gravel. In 1933 the barge transport company of J.J. Prior purchased the small sand and gravel business based at Fingringhoe, delivering to customers by river transport.

▼ Church Lane Stanway, a remaining cliff from the large quarry into Thames gravel and glacial outwash. One of the few sites where these gravels can be seen.

The adoption by Essex County Council of the Essex Minerals Local Plan in 2014 was a milestone in providing the county with the framework needed to ensure a steady supply of aggregate, supporting alternative sources of supply and managing mineral development within acceptable social and environmental limits.

Moulding sand for brass casting and iron moulding was formerly dug at Billericay from the Bagshot Sands. These sands are particularly fine and were very suitable for this purpose. Thanet Sand has been worked at Orsett Quarry, Linford, near Tilbury for the manufacture of building blocks. Silica sand is another significant mineral resource found in Essex. It is classified as an 'industrial sand' and its distinction from construction sand is based on its applications/uses and market specification. Silica sand contains a high proportion of silica in the form of quartz and has a narrow grain size distribution compared with other sand in Essex. The resources in Essex are at Ardleigh, north-east of Colchester, from a mixed resource. Industrial uses include glassmaking, foundry casting, ceramics, chemicals and water filtration.

▼ Sand and gravel aggregates at Colemans Farm Quarry, Little Braxted, where river terrace gravels and lake-bed sediments overlie the Witham buried valley. The underlying valley filling of blue clay provides a glimpse into the icy past. Gravel extraction at this site has provided an opportunity for research into the archaeology of flint artefacts and ice age mammal teeth and bones.

Beach recharge

In February 1969, the Tilbury Group based at Wivenhoe Pit won a contract to deliver over 26,000 tons of sand to build up Frinton's beach as part of an improvement to sea defences. There was local scepticism as to whether it would be washed away immediately, but the construction of new breakwaters stabilised the new beach material.

More recently, Holland Beach to the north of Clacton was recharged with gravel dredged from an offshore bank in a scheme to protect more than 3,000 homes and businesses for the next 100 years. For this, the physical modelling was carried out by an agency specialising in hydraulic research, which tested the design under varying conditions, including what would happen along the coastline in heavy storm conditions. This scheme is made up of 23 fishtail rock groynes and approximately 950,000 cubic metres of sand and shingle beach recharge extracted from a licensed offshore site 19 km (12 miles) east of Walton. In addition, more than a quarter of a million tonnes of larvikite rock were delivered in large blocks from Norway. After 349 trips to the offshore dredge site, 863,000 cubic metres of gravel was dredged and pumped onto the foreshore. The beaches were completed and fully opened to the public in 2015.

Two geological opportunities were opened up during and following this recharge scheme. From the recharge material, a wealth of Doggerland flint tools and mammal remains was found along the beach; and during the renovation of the cliffs, the sands and gravels of the ancient Thames and Medway rivers were revealed, enabling valuable research to be carried out (see Chapter 9 on the ice age).

Mammoth

Woolly Rhino

Horse

John Ratford collection

▼ A bonanza of Doggerland teeth (left) and worked flints from the recharged beach at Holland-on-Sea.

▲ Pumping offshore gravel to create a new protective beach at Holland-on-Sea and Clacton. Remarkably intact ice age bones, teeth and flint artefacts have been found on the beach here since the recharge was carried out; these provide valuable research material. *John Ratford*

Palaeolithic (Old Stone Age)

Palaeolithic (Old Stone Age)

Mesolithic (Middle Stone Age)

John Ratford collection

Development of water supplies in Essex

Early human settlements across Essex inevitably needed to be around or close to a supply of water. For thousands of years, people would have used rivers, streams, springs, ponds and lakes. Later, as populations grew, and farming developed, these supplies were not sufficient in either quantity or location. People then started to dig wells to access water from the rock layers beneath the soil. Mostly these layers are of water-bearing or waterlogged gravels, gravelly sands or 'head' deposits that had spread downslope during summer thaws of permafrost in cold stages of the ice age. This underground water supply is replenished by rainfall, mainly during the winter. The water is stored naturally within the rock, or it flows, seeps or trickles through the sediment by percolating downslope between sand grains and pebbles. Layers of rock that contain water that can flow through the rock are called aquifers and the natural rest-level of the underground water within the rock layer is called the water table. The supply of water from these underground sources is frequently referred to simply as groundwater.

When water is drawn from a well it is replenished by flow from the surrounding aquifer and, in turn, the aquifer is replenished by surface water and rainfall permeating through the rock. In many areas an increasing number of wells resulted in the aquifers being drawn down faster than they could be replenished. The water table was lowered and subsequently the shallow wells would run dry. Summer droughts would dry out many of these shallow wells and those that retained a summer flow became valuable sources in a drought-stricken community. To increase overall supply, wells were either re-dug to greater depths or boreholes were drilled and pumps installed. Clay augers – hand drills – were used for digging, turned by hand by two people. Extension pieces were added as the auger was driven down. Initially these deeper wells were pumped by hand, drawing water through a 25 to 75 mm (1 to 3 in) pipe, from depths down to about 16 m (50 ft). The pumps were hand-operated by a pump handle driving a piston with leather washers. By the end of the eighteenth century many villages had a pump, but only the affluent had a house pump, usually in the garden. The danger of pollution and diseases such as cholera was ever-present as the population grew and contamination became more prevalent with communal use. Some villages still have their former village pump in place, acting as a reminder of those good old days.

◄ An old domestic hand pump in a front garden along Main Road Broomfield.

140   STRAY NOTES ON ESSEX.

There is an account, dated 1791, of the sinking of a deep well at East Hanning-field parsonage, which, as it is not mentioned by Mr. Whitaker ("Geology of London and Part of the Thames Valley") may be usefully given here :

"It was begun June 21, 1790, and water, when the workmen, from such tedious labour, were at the moment of despair, was found May 7th, 1791. Thirty-nine thousand five hundred bricks were used, without cement, in lining this well, the soil of which, for the first 30 ft. was a fine, light brown, imperfect marl ; and though fossilists may ingeniously choose to discriminate the different strata, yet, except from shades of a deeper colour and firmer texture, occasionally, but slightly, mixed with a little sand and a few shells, the same soil, to a common eye, without more material variation, continued to 450 ft., where it was consolidated into so rocky a substance as to require the being broken through with the mattock. A bore then of 3 inches diameter and 15 feet in length, was tried, which soon, through a soft soil, slipped from the workman's hands and fell up to the handle. Water instantly appeared, and rose within the first hour 150 feet, and, after a very gradual rise, now stands at 347 feet, extremely soft and well flavoured. This source is supposed to supply the well at Battle's Bridge, about six miles further, and lower than Hanningfield, which is 336 feet in depth, and the water over-flows the brim. At Bicknacre Priory, 1¼ miles in descent from Hanningfield, is a well (nearly through neglect choked up) only 4 feet in depth."

It would appear from the details given that the London Clay was pierced through, and water obtained from sandy beds beneath, belonging to the Woolwich and Reading series. It is possible that the "fine light brown, imperfect marl" at the surface was the Boulder Clay, but if so, it is strange that the difference between it and the underlying London Clay escaped notice even from the "common eye" of the narrator. The rocky substance at a depth of 450 feet may have been at the base of the London Clay. A glance at the map of the Geological Survey shows East Hanningfield standing partly on London Clay, partly on Boulder Clay, so that the nature of the surface rock at the well remains uncertain. And there can be little doubt that the well at Battle's Bridge, which is from 40 to 50 feet above ordnance datum, while East Hanningfield is from 180 to 200 feet, is supplied from the same source. The height of the surface at Bicknacre Priory is about 150 feet.

From *Essex Naturalist*, Vol. 9 (1895),
Stray notes on Essex: the 'Gentleman's Magazine Library'.
Essex Field Club Archive

By the early 1800s a number of local authorities had taken over the operation of boreholes, particularly as they had to be drilled ever deeper to reach an adequate flow of good quality water. Typically, boreholes were drilled to depths of 35 to 50 m (100 to 150 ft) into the aquifers with bores of 100 to 150 mm (4 to 6 in) and the water was lifted by steam engine pumps. Later they were fitted with electric pumps. The mid-nineteenth century saw the formal development of water supplies with the formation of a number of private water companies across the country. In Essex two water companies were formed.

Much of Essex is clay country and the London Clay overlies the Chalk throughout most of the area. The clay is not permeable and does not act as an aquifer. The chalk beneath it is fractured and permeable and it acts as a large, deep aquifer, one of Britain's most important groundwater sources. The Essex Chalk aquifer is continuous with the north Kent Chalk and the Chilterns Chalk and is replenished by rainwater from both areas. Chalk had been quarried for some time in south Essex for use in cement manufacture, as a filler and as a soil conditioner for Essex clay land. There were two large companies operating opencast pits on the north bank of the Thames. As the pits were dug deeper, they encountered the water table in the chalk aquifer and the pits began to flood. To overcome this, pumps were installed to pump the water into the Thames. At that time workers lived close to their place of employment and alongside the pits were pit workers' houses. It was soon discovered that the

▼ Grays chalk quarry and waterworks in 1890.
Thurrock Museum

waste water from the pits was of a superior quality to the water from the local village pump and, as it was being pumped away, supplies were diverted and pumped directly to the workers' houses. This soon became recognised as a great benefit and there was strong demand to extend the supplies to local villages. The pit owners recognised the potential and in 1861 the South Essex Water Company was formed which eventually grew to supply a large area, from Grays to East Ham east to west and from the Thames to Brentwood south to north. In 1865 the Southend Water Company was formed to supply the areas of Southend, Billericay and Rochford. They also took over and operated 39 wells and boreholes in the area.

At the end of the nineteenth century the local authorities in the mid-Essex area developed their own water supplies, taking over existing wells and boreholes and developing new ones. Some of the borehole sites include Dagenham, Grays, Romford, Linford, Ilford, Roding, Stifford, Seven Kings, Galleywood Road Chelmsford, and Spital Road, Maldon. At some of the larger shafts, adits were dug, large tunnels radiating out from the bottom of the borehole into the Chalk aquifer to aid the supply of water into the borehole. These larger holes could be up to 4 m (12 ft) in diameter and contain several sets of pumps. Of these only Roding, Stifford and Linford are retained and these only operate when supplies from the other large treatment works are getting low. The others were closed down in the 1970s and 1980s and the sites sold off. In the Maldon area there was also a problem with natural fluoride in the water. Although fluoride is recognised as a benefit in dental health, excess fluoride results in a discoloration or mottling of the teeth. It was said at one time that you could recognise residents of Maldon by the colour of their teeth.

As demand for water grew in the developing area of south Essex, new supplies had to be found. In the 1920s water began to be extracted from rivers on an increasing scale, followed by the construction of reservoirs, Abberton in the 1930s and Hanningfield in the 1950s, followed by delivery from the Lee Valley reservoir complex in the 1960s. By this time the water supply in the southern half of Essex was predominately from surface water supplies and the few remaining groundwater sources were being phased out, with only 3% of the water supplied in south Essex coming from boreholes in dry years.

By the 1850s the lack of good drinking water had long been a complaint in north Essex. A borehole was dug at Harwich in 1854 by the forerunner of the Tendring Hundred Water Company, despite the failure of two previous wells to find water. Two years later the borehole had penetrated deep into the Chalk but had still not encountered a satisfactory supply of water. It was several more years before a good supply was found for the town. However, that borehole proved to be of great value to science: at the bottom, the hard basement rocks of Essex were revealed for the first time. Since that time, many water supply boreholes have been drilled across Essex, mainly into the Chalk aquifer. Waterworks, some of great architectural splendour, were built at well heads from where the high-quality water was pumped to the ever-increasing population, farms, horticulture and an increasing number of industries. Towards the end of the twentieth century, the Chalk aquifer was still an important source of the county's water supply. In the north-east of Essex, for instance, the Tendring Hundred Water Company was supplying 70% of its water from borehole sources.

Groundwater also comes into the Essex supply system from further away but, overall, the proportion of surface water supply across the county has increased considerably. There are several reasons for this change, one being maintenance of water quality. Due to the huge intensification in farming, a large quantity of nitrate fertilisers is used on the land. These have permeated down and there are now quite high levels in parts of the aquifers. Nitrates are not easy to remove and, in many areas of Essex, the concentrations of iron and fluoride in the groundwater are high. There are also financial reasons: it is more economic to have larger treatment centres than to have a number of dispersed small works and it is better to have a large, integrated distribution system to ensure continuity of supplies.

Furthermore, when aquifers are depleted it takes a long time for them to recover. Importantly, however, the confined Chalk groundwater in the Essex area – that is, the Chalk that is beneath the London Clay across most of Essex – is fully committed and no further water abstraction can be considered. At Langham, on the Suffolk border, the treatment works gets its supply from the River Stour but there are also four boreholes into the Chalk aquifer that are used in drought conditions when river flows are low. North-east Essex receives supplies from Ardleigh Reservoir but much of the rest of their supply is still from groundwater. Overall, the deep Chalk aquifer, once a major source of Essex water supplies, is depleted. It will take thousands of years to become completely replenished to its former water storage capacity.

▼ A plaque on the site of Dovercourt Spa, Harwich. *Gerald Lucy*

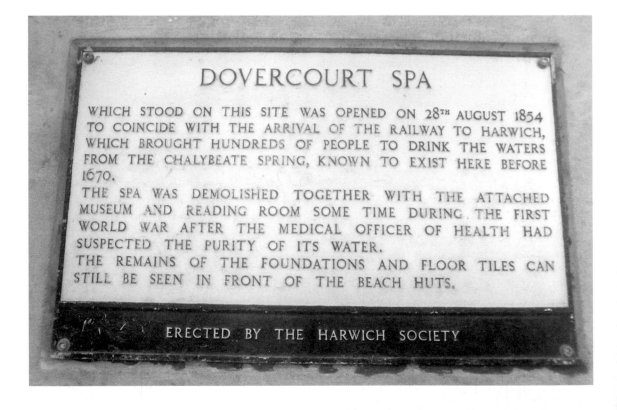

Essex spas

Water from wells and natural springs was, in parts of Essex, thought to have medicinal value and in the eighteenth and nineteenth centuries many towns became noted for their spas. The most famous of these was probably Hockley Spa, where a spring issues from the base of the Claygate Beds. The pump room is still in existence today. It was opened in 1843 and as trade developed, villas and a hotel were built for visitors. Christy and Thresh, in their book *A History of the Mineral Waters and Medicinal Springs of Essex* (1910), tell of a woman being employed to dispense the waters 'whose strong, healthy appearance visitors were led to believe was a result of the medicinal effects of the water'. However, despite this, the enterprise did not succeed and 'the public withheld its patronage and refused to be cured'.

One of the most successful Essex spas was at Dovercourt, where a fine spa house was constructed in 1854 overlooking the sea. The water was described as 'chalybeate', that is, containing a high concentration of iron compounds, and was considered to be of high therapeutic value despite, or perhaps because of, the awful taste. The chemical compounds in spa water vary depending on the rocks through which it passes; each source therefore possesses a unique character. There are records of at least 24 spas and medicinal wells in Essex; however, although some of these were very fashionable during the reign of Queen Victoria, the industry died out during the early years of last century.

▼ The former pump room of Hockley Spa. Built in 1842, the building can still be seen today. Hockley was perhaps the most famous of Essex spa towns in the nineteenth century. *Gerald Lucy*

13

The Naze beach near Walton. We can watch while the cliffs are removed and old rock is recycled to make new rock. The North Sea shoreline will advance and retreat through hundreds of thousands of years. The varied layers and time gaps in the future rocks of Essex will reflect a continued edgeland story. Eventually, the future of Essex will see the end of the ice age, the coming of new tectonic events with volcanoes and mountains, varied climates and new life forms. Massive changes will be every bit as normal – and as extraordinary to us – as the story of Essex past.

Essex Future

Now
+50,000
+100,000
years
► Anthropocene heating event and extinctions
Onset of next cold stage
Next glaciation
► Floods over Essex lowlands
Essex joined to Europe by Doggerland

The future of Essex rock

Climate, rock and recycling

The 500-million-year story of Essex rock reveals how dramatically the patterns of land and sea have changed. A map of the region of Britain has been unrecognisable throughout most of this time. In the last 65 million years Britain has split away from North America and has been left on the edge of the Europe-Asia continent. Only during this time has the map of the British Isles become vaguely recognisable, either as an oddly shaped peninsula of western Europe or, as now, a group of islands. Yet this map is ever-changing both in the short term and the long term; the future can never be really clear to us, on any timescale.

With our vision of the deep past derived from the rocks, we can at least try to peer forward into the geological future of Essex. It helps if we take time to contemplate what is happening to today's landscapes and rocks, and to delve into the wealth of evidence gathered over the last couple of hundred years, both from Essex and from many other areas, by many thousands of observers and scientists.

Throughout this book, we have shown that climate changes are normal. The effects of climate change in Essex have been extreme, not least during the last 3 million years of the ice age. Uniquely, human-induced changes are now, very suddenly, adding to the effects of any current natural changes. On a geological timescale, these added changes are instantaneous. That makes them catastrophic in nature. Such changes, however rapid or gradual, will be recorded in the Essex rocks of the future.

▼ Land is constantly worn down; rock is falling to pieces. Particles and fragments are carried away until they come to rest elsewhere. Where layers of recycled sediment become deeply buried, they eventually harden to become… new rock.

Erosion, rivers, alluvium and new Essex rock

Essex landscapes continue to change. In previous interglacial episodes, landscapes were subject to as much natural change as now. Now, however, human interference has modified nature, and at a greatly accelerating pace. What will happen to Essex in the near future depends upon very many things, not all of them predictable.

Some coastal changes are large and sudden by human standards, but others are slight and pass unnoticed by most people. All of these changes, and the current position of the coastline, are controlled by separate phenomena acting on different timescales. The coastline is affected by world sea-level rise, plus the rebound of the crust as the northern ice sheet has melted back, plus the slow tilt of the crust from the north-west as erosion and transport of the rocks of England continues. Added to these are changes in elevation caused by gradual adjustments in the underlying thick crust of the Anglo-Brabant Massif, some of which are now detectable by satellite measurements of minute surface adjustments. All of these changes combine in a complicated way, and all at different rates.

With the onset of the current Holocene interglacial stage, sea level started rising due to the melting of ice sheets after about 20,000 years ago. The effect of ice-unloading on the rocks beneath Britain is slow and this movement is also affected by the sea-level rise. In addition to all of these effects, the results of human activities such as water extraction, deforestation, land drainage and coastal defence constructions have geological consequences. With such complications of prediction, the future shapes and positions of the Essex coast cannot be certain.

Rising sea level

Currently there is no doubt that global sea levels are rising. Essex is particularly exposed to the effects of any increase in sea level. Artificial barriers and coastal protection measures can slow down marine and river erosion. However, they can also complicate the effects of tides, floods and storms, sometimes in unforeseen ways. Ultimately, at some time in the future, any large extent of additional protection is likely to become unaffordable and unsustainable. Inevitably, the magnitude and rates of change will eventually overwhelm the means to combat their effects. Storms, spring tides and high, sudden rainfall are factors that tend to overcome human effort. There will come a time, unpredictable though it may be, when some parts of the Essex coast and Thames Estuary will be catastrophically inundated.

Nature will tend to reassert the course of the Thames and other rivers across their natural flood-plains. Eventually, coastal marshes and saltings will be established according to natural, rather than artificial constraints – artificial coastal defences and sea walls tend to prevent the build-up of saltmarshes and extra land. Inland rivers will flood, erode the land and deposit alluvium according to natural river action and climate, maybe modified for some time to come by artificial embankments, weirs and flood defences. Yet ultimately these too will be eroded and corroded and transported away to sea. This will happen eventually whether the present climate is altering rapidly or slowly; it is just a matter of time. Essex will not stay constant; it will not remain as it is now. It never did stay constant. This book has sought to explain how we know. Uniquely, geological research into the past provides clues to future change.

Geology while you wait: a look into alluvium

Through the momentary timescale of modern human lifetimes, it can be difficult to understand that geological processes continue as they have done for countless millions of years. For instance, on a geological timescale, the surface of present-day Essex is eroding away quickly. The eroded residues of the present landscape surface, including soil, is called alluvium; this sediment is being removed by water right now. We can see it in layers of soil, clay-rich mud, silt, sand and pebble gravel in Essex streams, riverbeds, creeks, floodplains and estuaries. It is not always possible to make a clear distinction between alluvium deposits and deposits on near-shore saltings and mudflats, hill-slope deposits or soils. In Essex, these are all intermingled and boundaries on geological maps may be a 'best guess', particularly where these sediments are overlain by buildings and roads. Nevertheless, geological maps reveal alluvium along valleys where rivers convey Essex towards the Thames and North Sea. These deposits are commonly depicted in a pale cream colour.

You can stand and watch geology 'happening' in Essex. The land here is being moved, piece by piece, pebble by pebble, particle by particle, downhill, along rivers and ditches, from beaches and cliffs and into the sea, to be swept out onto the seabed or onto coastal marshland, eventually to form new rock layers. This is all part of the normal, unending geological recycling process.

Sediments are building up across the bed of the North Sea with a constant supply from the surrounding lands. Over the past 3 million years or so, including the time of the deposition of the Red Crag, layers of sediment have occasionally spread across wider areas around the North Sea, depending upon climate and sea level changes. The central North Sea, which is constantly sinking, is receiving the most complete rock record and this sedimentary build-up continues today.

▼ Concrete and larvikite pebbles are distributed by erosion and wave transport along the Naze beach. Geology and civilisation are moved around by natural erosion and transport.

◄ Pebble transport is gradual on our human timescale, yet these flints, already recycled three or four times in the past 60 million years, are on their way to a new home, in front of your very eyes, in Thorndon Country Park – as seen on the 'Pebble Walk'.

The land area of Essex, this edgeland, has a very occasional, incomplete rock record with many time gaps. In millions of years' time, it is possible that further 'Crag'-like sands will cover parts of Essex, occasional extensions of the build-up of rock layers of the North Sea area, whenever the sea level is high enough to cover present-day land areas. We are living in the midst of a continual 'Crag' sediment build-up, occasionally adding to Essex geology. We have merely a fleeting glimpse of this story lasting a few millions of years.

Invisible forces are at work deep beneath the surface of Essex. The movement of continents continues regardless of human activity. The timescales involved are both unimaginably protracted on the human scale and also catastrophically short: the slow movement of the Earth's mantle results in the gradual collision of continents, yet the earthquaking responses are only too apparent in human terms. The Colchester Earthquake of 1884 was just one of millions of such shakes. Although that was small by world standards, it had a lasting human impact. It is evidence of movement within the thick continental crust beneath Essex, a tiny piece of evidence for continuing flow in the Earth's mantle affecting the plate beneath us, including its top layer of crust. The movements of such earthquakes can eventually add up to changes in the way rocks are laid down on the surface and eroded away.

Rate of change: research and knowledge

A most profound feature of what is currently happening, as discerned from the record in the rocks, ocean sediments and ice layers, is that current climate change is acting very rapidly. It is probably far too rapid for the present-day complexity of human society to cope with effectively. Past populations in Essex have

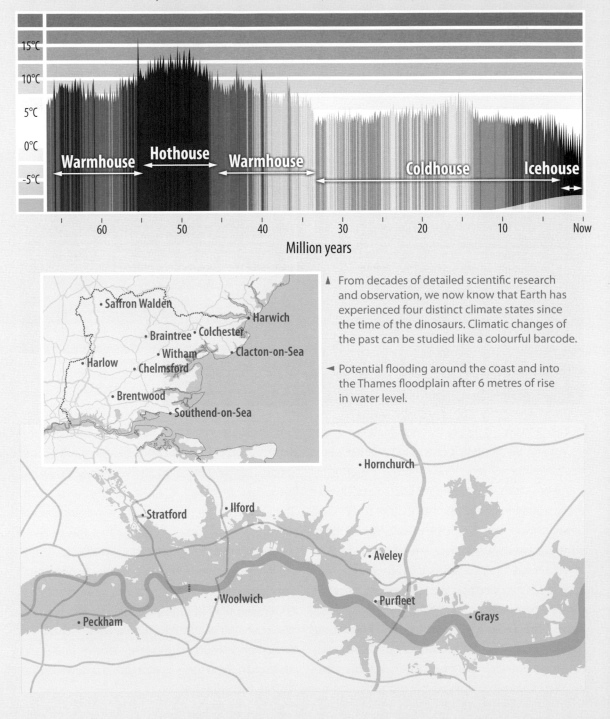

From decades of detailed scientific research and observation, we now know that Earth has experienced four distinct climate states since the time of the dinosaurs. Climatic changes of the past can be studied like a colourful barcode.

Potential flooding around the coast and into the Thames floodplain after 6 metres of rise in water level.

been small and self-reliant, even until well into the twentieth century. Villages were far less dependent upon a network of remote food supplies, for instance.

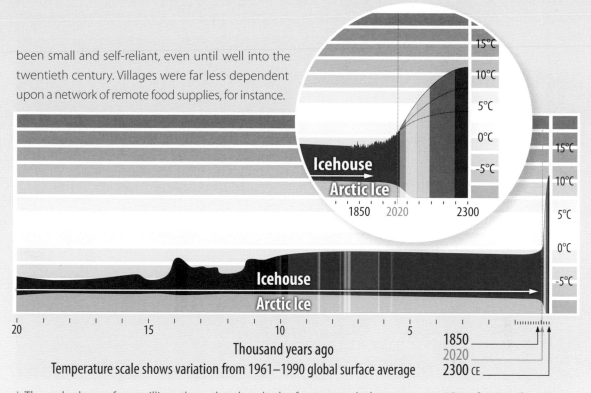

▲ The scale change from millions through to hundreds of years reveals the extreme rapidity of current human induced transformations. As a pointer to the future, the *rate* of change now far exceeds those of the warmest climate changes of the last 66 million years. The three dotted curves show alternative world climate predictions.
Derived from the Climatic variability of Cenozoic (CENOGRID) diagram compiled by Thomas Westerhold et al. 2020

Looking into evidence from the past within Essex rocks, scientists are continually piecing together the kinds of change, the rates of change and the reasons for the changes. Applying these to the present and future of Essex, to their likely effects upon land and sea and human life, is subject to the usual and varied levels of scientific knowledge and accuracy. The patterns of change during the ice age record for this county are still being discerned. They are valuable pieces of evidence to add to the global record of climate and ocean changes and, together, these are being used to aid prediction. Yet, there are few scientists working in this field of research, and very few are looking within Essex. Engineering and scientific research are not established as long-term political objectives; all too often they have been parts of short-term emergency reactions. True research has no predictable outcome; hence it is less likely to be appreciated and financed. Essex has the best record of evidence from the geologically recent past, yet its geological Sites of Special Scientific Interest are often inaccessible to inspection or otherwise completely lost to research. For our survival, the science of the future happens to be 'right now'.

It appears that plant and animal migration, from microbiology to large land and sea animals, now has far less time to adjust than at any time in the last few tens of millions of years. The effect on life will be unprecedented, not least in Essex, which remains, as shown from past times, an edgeland – a region particularly subject to small and rapid changes in climate, sea level and migration of plants and animals. Consequently, current changes are most apparent at the Essex seaside.

Sea defences and the future: the Naze example

As expected, the 2011 Crag Walk has prolonged the protection of the historic Naze Tower and of the main area of Red Crag geology, but at the expense of more rapid erosion further north. Predictably, it has also resulted in a build-up of vegetation across the prime geological research area. The erosion of the beach and cliff just north of the Crag Walk was immediate after 2011, with beach level lowered and now an apparently greater rate of removal by high spring storm tides of cliff rock debris further to the north. This process that removes already-mobilised London Clay from the base of the cliff is the most significant part of the mechanism that leads to rotational cliff failure of the higher section of cliff. The stable angle of rest for London Clay is only 8°, so it tends to keep moving. Meanwhile, the top of the Naze Hill continues to collapse and move seaward.

Little archaeological research has so far been carried out in the northern Naze. If the soils and deposits at the north part of the Naze are kept accessible for research, then they are more accessible to erosion. Time is of the essence. If the coast here is protected, then a build-up of vegetation may eventually prevent research until storms and erosion inevitably re-expose the beds. A short-term holding operation in the northern end of the Naze may arrest erosion for a time. Wooden groynes are relatively inexpensive and effective in allowing longshore drift to build up. This may be a good and quick option. From previous experience, it appears possible that heavy engineering could be a waste of energy, costly in time and money, and also could make things a lot worse overall. If nothing is done, then all the archaeological material will disappear quickly. Eventually it will all disappear whatever is done, but its loss could be slowed.

▲ Walton-on-the-Naze,
south and north
– a study in nature,
human interaction
and time.
Russell Wheeler

Seven-tonne blocks ►
of larvikite 'rip-rap'
imported from Norway
make up the seaward
face of the Crag Walk at
Walton-on-the-Naze.

▲ An eroding ancient land surface at the north end of the Naze shoreline.
Bill George

◄ The loss of land close to vulnerable sewage works at the northern part of the Naze.

It is particularly important that the total length of the Naze coast is considered as a whole rather than just concentrating on a particular part which, as seen elsewhere, has usually focused erosion into an adjacent section of coast – with unintended consequences.

The geological SSSI near the Tower is now almost totally lost as a valuable amateur resource and is now also fairly inaccessible for professional research, due to the overgrowth of vegetation on the stabilising cliff slope. One unfortunate consequence of the Crag Walk project is that any close in situ examination of the Red Crag, particularly by amateurs, is now greatly frowned upon, so sooner or later nature will obscure

it without further detailed study. It is a site with a rich history of study and discovery by amateur naturalists. Yet history, geology and archaeology can continue to provide evidence to illustrate what makes the Naze unique.

A hugely significant factor in the future of the Naze area is the proper appreciation of the different timescales involved. Yet this is very rarely considered and thus not appreciated for its overriding importance in assessing the impacts of human actions and the various interactions with nature. Future policy may be to 'let nature take its course'. This is inevitable eventually, whatever happens, because the means will

▲ Land on the move as cracks open out at the Naze clifftop. *Pete Jarvis*

become unavailable as time goes by. Inland, behind the Naze, the Walton backwaters will evolve, the sea will get there, and humans will either adapt or suffer disaster. Planning ahead for inevitable and surprisingly rapid change would seem to be a rational option, but now with some urgency. Time, as usual, will tell. Time is a key for wildlife too: it will tend towards looking after itself, without need of 'management' – or indeed any humans at all.

We find out about the future from the past. In considering the future, there may be no simple, single solution for humans. For nature, however, there could ultimately be a less drastic outcome.

Essex soil and the future

Civilisation is unsustainable without soil. If soil becomes sterile, or is removed by wind and water, then human society cannot survive. The present soils of Essex are not the soils that would have been found across the land had humans been absent throughout this Holocene interglacial. The interglacial temperate forests here have been almost entirely chopped down over the past few thousand years. Clearings in burnt wildwood were tilled and used as productive land until they were depleted. Then the families would move on and clear another section until that soil was depleted. Such practice had a finite progression throughout the forested area of the Essex countryside. Wildlife and wild plants exist within an interwoven ecological web, of which 'natural' soil is a living and extremely complex part. Agriculture and horticulture have stimulated an ever-increasing human population, epitomised by the current situation in Essex. They are overwhelming ecological intrusions that have destabilised the ecosystem of the current Holocene interglacial stage of the ice age.

The soils within most areas of Essex farmland are now maintained for high-yield food production, following repeated revolutions in agricultural practice to cater for the fast-growing human population. Essex is also crowded with human infrastructure, diminishing the farmland needed to support it: land is covered and made sterile. Farming and horticulture have increasingly used agrochemical input to maintain soil as an artificially supported growing medium. However, agrochemical manufacture is highly energy-intensive and chemicals such as phosphate for fertiliser have to be imported. Unfortunately for current human populations, it has taken only a few decades to exhaust soil of its natural recycling potential. It becomes degraded and easily lost to water and wind. Then it takes hundreds of years to get back to equilibrium involving geology, biology and natural input of dead matter for recycling into living matter. This is the basis of our fear for the future. Some Essex farmers are questioning the logic and changing farming methods. Farmland depends more than ever upon the application of policies founded on scientific observation plus farming experience from the past such as diverse crop rotations. With no-till farming, cover crops and crop residue protect the vital, natural organic content and allow rain into the soil without erosion. Carbon is absorbed from the atmosphere and not lost during ploughing, and productivity increases – a dual benefit. Essex society, food and wildlife depend upon the soils and sediments of the county, their stories of the past, their natural clues to the future.

Farmland erosion. ►
Wikimedia Commons/Dr Donal Mullan

▲ The Anthropocene? Evidence of human activity at the northern end of the Naze.

Anthropocene

Scientists have been discussing a new geological Era, the Anthropocene, to describe the evidence of the time of human influence on geological processes, including sedimentation. It now seems likely that, rather than a protracted era of change, the Anthropocene may be recorded as an extremely abrupt transformation – a 'spike' of change to a new equilibrium – in the sedimentary and temperature record of the planet within the immensity of future geological time. The current warming trend is more sudden than previous global heating events recorded from evidence gathered for the ice age – and even from previous global heating events such as the Palaeocene-Eocene Thermal Maximum event some 56 million years ago. Such a rapid change is catastrophic for the natural environment – causing the onset of a mass extinction event – and also for the human civilisation that has conceived the Anthropocene Era. Further study and application of geology and the other Earth sciences can only help prevent the most far-reaching effects of this abrupt transformation upon natural ecosystems and human life.

The next glaciation

On a geological timescale, we are still in the Quaternary ice age. Glaciers cover about 10% of the Earth's land surface, but they are shrinking rapidly across most parts of the world. In the European Alps atmospheric warming combined with decreased snowfall has led to a 54% loss of ice area since 1850 and this has been most pronounced during the last few decades. Current projections suggest that around 10% of the 2003 European Alps ice area will remain by 2100. There is evidence that the previous warmest period of the present interglacial was 5,000–7,000 years ago when the sea, which had been rising since the last glaciation, again flooded the lowlands of Doggerland to make Britain an island within the European continental edge. Now, the climate is warming again, but very much more rapidly than before.

Carbon dioxide emissions resulting from modern human activities now affect the natural oscillation in world temperatures. The planet has already warmed by more than 1°C through the past 200 years, which compares with a rise of only 4°C at the end of the last glacial stage. In 2013, the daily mean concentration of carbon dioxide in the atmosphere surpassed 400 parts per million (ppm) and by 2020 was over 415 ppm. Research indicates that carbon dioxide has not reached this level within the last 2 to 4 million years – and never so suddenly. If rates of emissions continue to rise, concentrations of greenhouse gases in the atmosphere will have large effects upon sea level, wildlife, agriculture and population in Essex – and throughout the world.

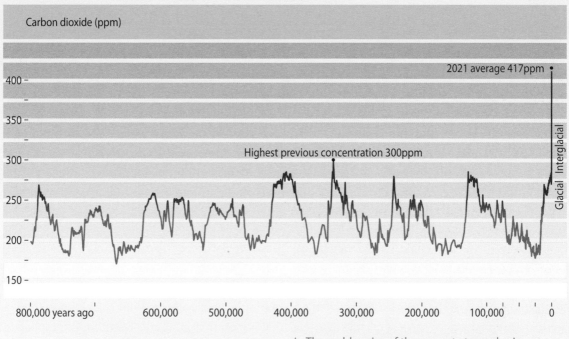

▲ The sudden rise of the current atmospheric CO_2 level – a depiction of catastrophic change.

Prediction based on orbital variations – the Milankovitch cycles (see Chapter 9) – indicates a possible onset of colder conditions in around 50,000 years from now, although this onset would likely be modified by the current anthropogenic atmospheric heating. It may even be prevented altogether if the emission of greenhouse gases escalates out of control, leading to a 'runaway' greenhouse effect until the planet becomes uninhabitable by any higher life-forms – much like the planet Venus.

Thorough scientific observations, study, knowledge and appreciation of the workings of this planet, its rocks, life, atmosphere and hydrosphere are literally vital, not merely to human life, but to all life. Geologists have a very important role to play in interpreting the evidence from the rocks. In particular, the record of the sediments throughout Essex contains many clues to the way our climate has changed. Essex research results, added to those of many results elsewhere, could provide crucial evidence to support attempts to reduce the impacts of the current enhanced global heating – given the opportunity.

The pattern of gradual cooling though past interglacials has shown a series of increasingly severe stadials (cold snaps) alternating with warmer spells, each lasting a few thousand years or so, until the greatest cold of a full glaciation stage sets in. It is possible, therefore, that Essex and its edgeland coastal areas could be affected once more by large oscillations in sea level and climate a lot sooner than 50,000 years from now, culminating in the eventual onset of full glaciation, with a consequent drying-out of the seabed to form new plains and lakes of the next Doggerland. Perhaps there will be a renewed flow of migratory species to and from the lands of Essex and across what is currently a landscape beneath the waves. However, there might be fewer species of large mammal grazing the plains following the onslaughts of human activity during this current interglacial.

Beyond the ice

The more distant future of Essex is much harder to discern. With the large amount of information on plate tectonics that has been gathered over the past 50 years, a trend for our area can be suggested. Inevitably it will involve huge changes, just as we have seen from the past few hundred million years – ever since Essex was founded in the volcanoes and mud around the edge of a supercontinent close to the South Pole.

Essex is currently part of the continental crust area of Europe and Asia and is not far from the Atlantic Ocean crust. Depending upon global climate and sea levels, Britain in this ice age is variously a peninsula or a set of islands within this continental shelf area. However, our piece of the continent is subject to both continental drift and continental collision. So where will Essex be among all of this activity in many tens of millions of years' time, beyond this ice age? There are various predictions, based on present and likely movements of continental and oceanic crust.

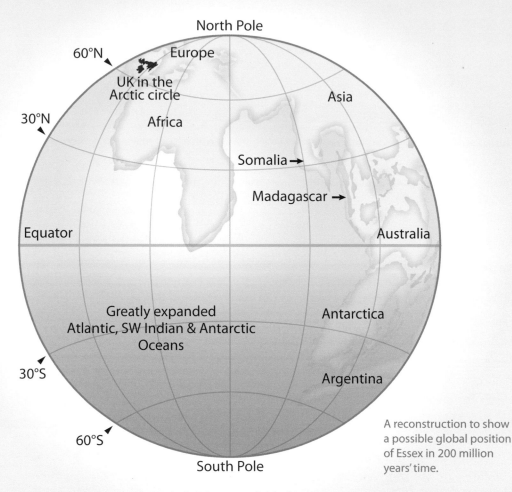

North Pole

60°N ➤   Europe

UK in the
Arctic circle

Asia

30°N ➤   Africa

Somalia ➤

Madagascar ➤

Equator   Australia

Greatly expanded
Atlantic, SW Indian & Antarctic
Oceans

Antarctica

30°S ➤

Argentina

60°S ➤

South Pole

A reconstruction to show
a possible global position
of Essex in 200 million
years' time.

Britain is currently drifting northwards, having passed through various climatic zones over the past few hundred million years – adding to the wonderful variety of sediments seen in Britain's existing rock layers. Within the coming few hundred million years, Essex is likely to continue drifting further to the north. A few millimetres of movement each year results in many kilometres of continental drift over millions of years. Meanwhile, the African continent continues to push against the European continent, closing the Mediterranean and continuing the squeeze of the great Alpine mountain range. The Alpine chain will become larger and more extensive as time goes by, until the complex merger is done and, maybe, further fragments of continent open out and create new ocean crust in this unending slow dance.

Added to all that, it is likely that there will come a time when the Atlantic Ocean starts to close again, as the Pacific Ocean crust continues to widen. It is probable that a new subduction zone will form at either or both Atlantic margins. Such a zone along the edge of Europe would have a profound effect on the Essex area, bringing a chain of volcanic mountains or an island arc similar to that of present-day Japan, until the American continent merges with us once again. Or Essex could escape and drift further north and become a quiet, geologically stable area, perhaps adjacent to Siberia. Either way, by that time, almost every trace of human civilisation would be either completely eroded away or deeply buried.

Learning from the past: the key to the future

Although such far-future predictions cannot directly affect
Essex today, or our decisions, an appreciation of and research into
the future provides a perspective on how things have developed
to get us where we are now. That, in turn, should help with
decision-making in the shorter term. For the edgelands of Essex,
such decisions are crucial.

The more we can learn from detailed geological investigations,
notably across Essex with its ice age and recent sediments and
landforms, then the more we can attempt to become informed
and prepared for practical short-term action.

If we are to understand what is happening to us now
and to attempt to look after the future of the natural world,
we need to understand the past that has brought us
to this point.

Shakespeare put it quite simply in *The Tempest*:
'What's past is prologue.'

A selection of sites and views of Essex geology and landscape

These sites are publicly accessible at the time of publication (2022) or can be viewed from public footpaths. Ordnance Survey map references are given for each location. The inclusion of a site in the list does not necessarily imply a right of access. Please look after these areas and do your best not to disturb property or wildlife. Sites for standing stones and boulders are just a selection from many across the county.

Detailed descriptions of many sites of geological interest and importance are provided on the Essex Field Club website. Go to: A–Z Geological Site Index.

ALPHAMSTONE. *Alphamstone Churchyard*
(TL 878 354), Braintree district.
There are eleven sarsen stones here, eight in the churchyard, two by the road and one in the church. There are probably more buried in the ground locally. These are erratic boulders that have been gathered on this knoll in the distant past.

ALRESFORD. *Cockaynes Wood Nature Reserve*
(TM 056 218), Tendring District.
A former gravel pit with exposures of Wivenhoe Gravel, part of the bed of the ice-age River Thames half a million years ago. It was deposited just upstream of the point where the Thames joined a former course of the River Medway.

ALTHORNE. *The Cliff SSSI*
(TQ 921 968), Maldon district.
A cliff on the outer bend of the River Crouch near Burnham-on-Crouch is being eroded. Here, London Clay fossils have been found in the beach shingle including fossil bird bones and numerous shark teeth. There has been extensive collecting since the site was first discovered in the early 1970s and finding fossils here today requires good eyesight and a lot of patience. Transparent crystals of selenite (a variety of gypsum) can also occasionally be found.

ARKESDEN. *Arkesden War Memorial*
(TL 4821 3456), Uttlesford district.
The war memorial in St. Mary's churchyard consists of a very large boulder of puddingstone 1.7 × 1.7 × 0.8 metres in size. This boulder was apparently brought here from somewhere on the Wood Hall estate, probably less than a mile to the south. In the bed of the stream (Wicken Water) by the road bridge in the centre of the village are a number of large erratic boulders.

AUDLEY END. *Audley End House Septarian nodule*
(TL 5217 3831), Uttlesford district.
On display in the Stable Block is a fine, large septarian nodule (165 × 115 × 30 cm in size) cut in half to reveal the internal calcite-lined cracks or 'septa'. This nodule was no doubt collected locally and was part of the natural history collection acquired by the 4th Lord Braybrooke in the nineteenth century.

BEAUMONT. *Beaumont Quay Limekiln*
(TM 190 240), Tendring district.
The circular brick limekiln at Beaumont Quay is the only complete limekiln surviving in Essex, dating from around 1869–70. It had fallen into disuse by the early 1920s. Limekilns were often built in harbours and wharfs where chalk and coal for the kiln could be brought in by sea. Owned by Essex County Council and publicly accessible.

BLACK NOTLEY. Friar's Farm Boundary Stone
(TL 7427 1970), Braintree District.
A sarsen stone that is described by English Heritage as a rare survival of an inscribed and dated boundary stone dating back to the seventeenth century. Sarsen stones used for specific purposes in historical times are rare and this is a splendid example.

BRENTWOOD. *Thorndon Country Park North*
(TQ 604 915), Brentwood district.
Thorndon Country Park has some interesting landforms which enable the geology to be appreciated. To the south a spread of glacial gravel is dissected by several streams, which have cut through the Claygate Beds exposing the underlying London Clay. Glacial gravel is well exposed in a gravel cliff at the east end of the park with a signboard explaining the geology. A geological 'Pebble Walk' trail guide is available.

CHAFFORD HUNDRED. *Chafford Gorges Nature Park*
(TQ 599 793), Thurrock district.
Chafford Gorges Nature Park is the finest area for geology in south Essex. Spectacular Chalk cliffs can be seen, a legacy of quarrying for the Portland Cement industry. The Chalk is overlain by Thanet Sand and gravels from former routes of the Thames during the ice age. There are also several fine sarsen stones around the rim of Grays Gorge. The park consists of seven geological sites (Grays Gorge, Lion Gorge, Lion Pit Tramway Cutting SSSI, Millwood Sand Cliff, Sandmartin Cliff, Warren Gorge and Wouldham Cliff), looked after by Essex Wildlife Trust. Lion Pit Tramway Cutting SSSI is one of the most important of the county's SSSIs, yielding evidence of human occupation on the banks of the Thames 200,000 years ago. The visitor centre has views overlooking Warren Gorge. A geological trail guide is available.

CHELMSFORD. *Chelmsford Museum Puddingstone*
(TL 7025 0558), Chelmsford district.
Large boulder of puddingstone, with a signboard to tell its story, in the ornamental garden near the entrance to Chelmsford City Museum and café in Oaklands Park.

COLCHESTER. *Lion Walk United Reformed Church*
(TL 9971 2506), Colchester district.
The top of the church spire fell into the street during the 1884 earthquake and the Illustrated London News at the time published an impressive engraving of the occurrence with people fleeing from the scene. The church has since been demolished but the tower (with its rebuilt spire-top) was spared and it remains as a prominent landmark in the town shopping centre. A plaque commemorates the event.

EAST MERSEA. *Cudmore Grove Cliffs and Foreshore*
(TM 068 146), Colchester district.
The cliffs at Cudmore Grove Country Park provide superb exposures of gravels laid down by the ancient River Blackwater during a glacial period 300,000 years ago. Riverbed sediments at beach level sometimes yield fossils, including mammal bones, that indicate that they were deposited during an interglacial period. Also exposed on the foreshore are deposits from a more recent interglacial period, the Ipswichian interglacial (120,000 years old), and known as the 'hippo site' due to the presence of hippopotamus bones.

EPPING. *Wintry Wood Brick Pit*
(TL 469 035), Epping Forest district.
Disused brick pit in Wintry Wood adjacent to Brickfield Business Centre (formerly Wintry Park Brick & Tile Works). Steep banks have Anglian glacial till ('chalky boulder clay') visible in tree roots. This pit is a rare example in Essex where direct evidence of the ice sheet can be seen. The pit is in the Lower Forest (part of Epping Forest). Layby for parking nearby.

EPPING FOREST. *High Beach Viewpoint*
(TQ 411 981), Epping Forest district.
The high ground around the King's Oak inn is capped with High-Level Pebble Gravel (Stanmore Gravel), which is likely to have been deposited by a tributary flowing north from the area of Kent, to join the ancestral Thames across north Essex and Suffolk (gravels of similar composition occur at the top of Shooters Hill on the south side of the Thames). Well-rounded flint pebbles from this gravel are revealed in the many footpaths. Beneath this gravel is Bagshot Sand from a delta that spread across Essex 50 million years ago. This fine yellow sand is visible on the steeply sloping paths to the north-west of the inn.

FINCHINGFIELD. *Finchingfield Boulder*
(TL 6849 3290), Braintree district.
A splendid boulder of basalt volcanic rock 85 centimetres long lies by the roadside on the left-hand side a few metres from the village green on the road north out of the village. This boulder was probably transported south from Northumberland or Scotland by the Anglian ice sheet. Large erratic boulders of basalt are very rare in Essex.

FINGRINGHOE. *Fingringhoe Wick Nature Reserve*
(TM 045 195), Colchester district.
Essex Wildlife Trust nature reserve. Fingringhoe Wick was a working gravel quarry until 1959. Visible in many places are mounds and banks of glacial sand and gravel (known locally as Upper St Osyth Gravel) which was deposited some 450,000 years ago by torrents of meltwater issuing from the Anglian ice sheet, the edge of which was then situated only 12 kilometres west of here. The gravel therefore provides evidence of an exceptionally cold period in the ice age. A permanent vertical section through the gravel exists in the centre of the reserve.

GESTINGTHORPE. *Nether Hall Farm Sarsen Stones*
(TL 809 393), Braintree district.
Nine sarsen stones on roadside by farm entrance, the largest 2.4 metres long. The area around Gestingthorpe is notable for the large number of sarsen stones that can be found at road junctions and by farm gates, all recovered from local fields by farm workers.

GREAT WAKERING. *Star Lane Pits*
(TQ 939 872), Rochford district.
Low cliffs of yellowish brickearth (loess) can be seen adjacent to footpaths in this pit in a wildlife reserve. Loess is a fine silt deposited during huge dust-storms carried from the cold, dry land around the edge of an ice sheet. It was possibly deposited during the coldest part of the most recent glaciation of Britain, by about 20,000 years ago.

HADLEIGH. *Hadleigh Castle Landslip*
(TQ 810 860), Castle Point district.
The most impressive London Clay landslip in Essex. The severe effects on the castle are obvious. Further, larger slipping in the field and footpath to the east, towards Leigh-on-Sea, are seen from the eastern edge of the castle.

HADLEIGH. *Hadleigh Country Park*
(TQ 799 868), Castle Point district.
Undulating landscape of London Clay, overlain by Claygate Beds and rounded hills of Bagshot Sand. An exposure in the Bagshot Sand delta was created in one hill in 2016, close to the site of British Geological Survey's 1973 Hadleigh borehole. The small cliff of yellow sand and a signboard are on the geological trail around the park, with a trail guide available at The Hub. Here also, in the topsoil, fragments of Kentish chert looking like small lumps of toffee tell of the former course of the ancient River Medway across this area. Fine views across meanders in local rivers on the floodplain and across the Thames to Kent and the Chalk of the North Downs.

HARWICH. *Harwich Foreshore SSSI*
(TM 263 320), Tendring district.
This locality has the best exposure of the Harwich Stone Band, the most distinctive of the cementstone bands in the Harwich Formation at the base of the London Clay. It contains volcanic ash from explosive volcanic eruptions in Greenland as the North Atlantic started opening out during Eocene times, some 55 million years ago. The stone band makes this part of the coast the only naturally occurring rocky shore along the entire distance between Norfolk and Kent and may be one reason for the existence of the Harwich peninsula. The foreshore is also of prime importance for London Clay fossils, particularly for fossil fruits and seeds from the Eocene rainforest. Also found are fossil shark teeth amongst the beach shingle.

HAVERING-ATTE-BOWER. *Bedfords Park*
(TQ 5200 9222), Havering London Borough (Essex vice-county).
A boulder of dolerite from the Whin Sill lies on the terrace of the Essex Wildlife Trust visitor centre, with a signboard on the other side of the wall. The boulder was transported within the Anglian ice sheet from the north of England. The Whin Sill is a feature across the landscape upon which the Hadrian's Wall was built. This boulder was evidently caught in the diverted Thames river flowing around the edge of the melting ice sheet to the south of here, perhaps in an ice floe, and came to rest in the gravel riverbed. It was found in a gravel pit at Marks Gate 450,000 years later and kindly donated by Brett Aggregates. From this area of the park there is a fine view of the 'staircase' of lower Thames terraces and across to the Chalk of the North Downs of Kent.

INGATESTONE. *Ingatestone Boulders*
(TQ 6511 9967 and TQ 6511 9959), Brentwood district.
Two well-known sarsen stones protect the junction of the High Street and Fryerning Lane, the larger one a metre tall.

Another stone is situated a short distance away by the south door of St Edmund and St Mary Parish Church which has one of the earliest brick-built towers in Essex, plus a north wall of ferricrete and Roman brick.

LANGDON HILLS. *Langdon Hills Country Park*
(TQ 683 866), Basildon district.
London Clay and Claygate Beds with Bagshot Sand capping the highest ground. This in turn is capped by High-Level Gravel (Stanmore Gravel), an ancient bed of a river flowing from the south. There are sunken lanes up the steep hill to the west where the old road cuts into the Bagshot Sand layer. Disused sand and gravel pits are seen in the woods and small toffee-like fragments of Kentish chert are common amongst the flint pebbles in the footpaths. Spectacular views across the Land of the Fanns (Bulphan Fen) and the Thames Estuary and most of London. Part of the site is in Thurrock District.

LEXDEN. *Lexden Springs Nature Reserve*
(TL 973 253), Colchester district.
In Spring Lane, off Lexden Road, is a patch of ancient meadowland which is generally rich in wildflowers. Here a natural spring issues from the junction of the Kesgrave Sands and Gravels and the underlying London Clay.

LITTLE WALTHAM. *Channels Sarsen and Puddingstone*
(TL 7238 1118), Chelmsford District.
By the entrance to the Channels venue is a large boulder of puddingstone on a mound of grass by the roadside. It is one of the largest puddingstone boulders in Essex, measuring more than 2 by 1 metres. Close by are boulders of sarsen, including one large piece that has a trace fossil trackway along its front surface, plus the root holes left from plants that grew in the sandy soil that became transformed into this tough sarsen quartzite. These were found in the former Channels gravel pit.

LITTLE WALTHAM. *Channels Glacial Till Wall*
(TL 7216 1100), Chelmsford District.
Within a new housing development, a public footpath offers a view of the only easily seen exposure, anywhere in Essex, of glacial till, the Chalky Boulder Clay left by the Anglian ice sheet that covered the Thames gravel in this area. At the time of writing (2022) a signboard is proposed to be set next to this remaining quarry face of the former gravel workings.

MAYLANDSEA. *Foreshore*
(TL 907 035), Maldon district.
The foreshore on the east bank of Lawling Creek, north of Maylandsea, has yielded large numbers of London Clay fossils, particularly crustacean remains such as lobsters. The SSSI citation states that this section of 'soft rock' coastline is of national geological importance. The site demonstrates the maximum extent of the London Clay sea, with fossils indicating a deep-water fish fauna.

MISTLEY. *High Street Cobble Wall*
(TM 117 318), Tendring district.
On the south side of Mistley High Street, just east of the post office, is a brick wall with a panel composed entirely of rounded cobbles. Here can be seen numerous 'exotic' rock types such as granite, dolerite and gneiss and the distinctive Norwegian rock known as 'rhomb porphyry', of which there are at least eight examples here. The high number of the Scandinavian rocks in such a small section of wall is extremely unusual and, situated opposite the Baltic Quay on the adjacent Stour, and 'Ballast Hill' within the river, it is more than likely that these cobbles are of ships' ballast from maritime trading 150 to 200 years ago. The wall is a geological museum of delight.

MISTLEY HEATH. *Furze Hill Gravel Pit*
(TM 122 309), Tendring district.
The disused gravel pit at Furze Hill, adjacent to the long-distance Essex Way footpath, has exposures of Waldringfield Gravel. This is the oldest deposit in Essex from the former River Thames, at least 650,000 years old. Gravel seen in small excavations along the paths and in roots of fallen trees contains 'exotic' rocks that have been carried some distance by the Thames, such as small pebbles of pure white vein quartz possibly from North Wales and rare black-and-white tourmaline quartz pebbles from Cornwall, as well as numerous flints of all colours.

NAVESTOCK. *Millennium Stone*
(TQ 5460 9613), Brentwood district.
A giant boulder of ferricrete sits on a concrete plinth by the road. Excavated from a nearby field, it is 2.1 by 1.5 metres in size and was placed here in 2000 to mark the Millennium. It is mistakenly inscribed as being a 'puddingstone'.

NEWPORT. *The Leper Stone*
(TL 5199 3496), Uttlesford district.
A large coarse-grained sarsen stone 1.7 by 1.2 metres in size known as the Leper Stone sits upright on the grass road verge at the north entrance to the village. This is the best-known erratic boulder in north Essex. Adjacent to

this is a wall constructed largely of blocks of clunch, a hard variety of chalk formerly used for building.

PURFLEET. *Purfleet Submerged Forest*
(TQ 5445 7871), Thurrock district.
Part of a submerged forest, between 5,000 and 6,000 years old, consisting of fallen tree trunks and roots, with some rare leaves, is exposed on the Thames foreshore, at low tide only. This site and other submerged forests along the Thames (e.g. at nearby Rainham) have been studied for their importance in the interpretation of sea-level changes since the end of glacial conditions 11,700 years ago.

SAFFRON WALDEN. *The Gibson Boulders*
(TL 5369 3817), Uttlesford district.
At the junction of Gibson Gardens and Margaret Way is a mound of grass and trees containing at least 25 glacial erratic boulders of varying sizes. At least ten different rock types are seen here. The largest is a slab of colourful puddingstone 1.2 metres long. The site also has great historical interest: the Gibson Gardens estate was built on land that was formerly the gardens owned by George Stacey Gibson (1818–1883), naturalist, who had a great interest in geology. An 1877 map shows this mound to be the site of his summer house and it seems certain that he accumulated these boulders in his garden. They were almost certainly gathered from the farmland that he owned in the vicinity.

ST OSYTH. *Colne Point Shingle Spit*
(TM 109 123), Tendring district.
Colne Point is the best example in Essex of a shingle spit. The spit is 4 kilometres long and is nearly all that remains of a much larger area that existed in the nineteenth century but has now mostly been developed by the seaside holiday industry. It is of great interest for studying the movement of shingle and the development of shingle structures. It is an Essex Wildlife Trust reserve. Day permits to visit are available from the Trust.

ST OSYTH. *St Osyth Priory Gatehouse*
(TM 121 157), Tendring district.
Erected in 1481, the battlemented gatehouse of St Osyth Priory is one of the finest examples in Britain of the use of flint 'flushwork' – the name given to the technique of setting 'knapped' flints (skilfully worked to produce a flat face) into a wall, often in intricate patterns alongside another stone

ST OSYTH. *St Osyth Priory Gatehouse*
(TM 121 157), Tendring district.
Erected in 1481, the battlemented gatehouse of St Osyth Priory is one of the finest examples in Britain of the use of flint 'flushwork' – the name given to the technique of setting 'knapped' flints (skilfully worked to produce a flat face) into a wall, often in intricate patterns alongside another stone such as limestone. St Osyth's Priory is privately owned and no longer open to the public but the gatehouse can be viewed from the green open space between the gatehouse and the road where there is also car parking. The adjacent Priory boundary wall along to the village crossroads is largely faced with excellent examples of buff-coloured cementstone, probably from Harwich, together with black flints.

SOUTH OCKENDON. *Davy Down Sarsen Stone*
(TQ 592 800), Thurrock district.
Davy Down Riverside Park forms part of the Mar Dyke Valley, a delightful valley with steep, wooded sides that can be seen from the M25 motorway. The valley runs along the northern side of the Purfleet ridge between Purfleet and Grays. Also of interest is the former waterworks containing old pumping engines, in front of which is a remarkable sarsen stone – one of the finest examples in Essex. The sarsen is 1.6 metres square and has fine mammillated surfaces.

STANWAY. *Church Lane Gravel Cliff*
(TL 945 238), Colchester District.
The disused gravel pits south of Church Lane have mostly been infilled and much of the land is now occupied by housing. However, a steep cliff of Thames gravel with glacial outwash gravel above is preserved south of Church Lane just west of the new Stanway Western Bypass. This cliff is a rare survival.

TOLLESBURY. *Woodrolfe Creek Saltmarsh*
(TL 969 105), Maldon district.
The best and safest place to view saltmarsh in Essex, with its creeks and tidal flats, is at Tollesbury. Here the marsh is criss-crossed by paths to enable boat owners to reach their vessels. This shows where new geological layers are being created.

UGLEY GREEN. *Ugley Green Puddingstone*
(TL 524 271), Uttlesford district.
Beside the green, next to the village pump, is a fine, rounded and colourful boulder of Hertfordshire puddingstone 1.2 metres long.

WALTON-ON-THE-NAZE. *The Naze Cliffs SSSI*
(TM 266 235), Tendring district.
The finest geological site in Essex. Classic cliff section in London Clay, Red Crag, brickearth and Thames Gravel (Cooks Green Gravel). Volcanic ash bands and faults in London Clay. Classic rotational landslips. Diverse fauna of fossils from Red Crag and London Clay. Site of international importance. A geology trail guide is available from GeoEssex and Essex Wildlife Trust. The tower contains a showcase of geological and archaeological finds and a GeoEssex display of beach finds is seen in a showcase within the EWT Centre nearby. The all-round views – for a small charge – of the land and seascapes from the top of the tower are truly stunning.

WENDENS AMBO. *Wenden Place Boundary Wall*
(TL 512 363), Uttlesford district.
On the bend of the main road opposite the church is a high, ancient wall, which is remarkable for the variety of local rocks used in its construction, including many large boulders. The largest is a puddingstone 1.4 metres long. The wall is a Grade 2-listed building.

WHITE NOTLEY. *White Notley Puddingstone*
(TL 7880 1772), Braintree district.
By a cottage gate is a fine, colourful boulder of Hertfordshire puddingstone 1.1 metres long. This boulder has been referred to in articles and books more often than any other puddingstone in Essex.

WRABNESS. *Wrabness London Clay Cliffs and Foreshore*
(TM 172 323), Tendring district.
The London Clay cliffs on the River Stour at Wrabness are the highest vertical cliffs in Essex and consist of the upper part of the Eocene Harwich Formation and the lower few metres of the Walton Member of the London Clay. They provide the best onshore exposure of the Harwich Formation. These cliffs are of particular interest because they contain a complete sequence of bands of volcanic ash, which originated from volcanoes in Greenland 55 million years ago. These ash bands are present above the Harwich Stone Band which is very well seen along the top of the beach. The site has also yielded an important fossil flora preserved in concretions, as well as mammal bones and worked flints from interglacial deposits further along the shore.

Some geological collections and displays of geological interest

Centre for Biodiversity and Geodiversity
The Green Centre, Wat Tyler Country Park,
Pitsea Hall Lane, Pitsea, Basildon SS16 4UH
The Essex Field Club's Centre for Biodiversity and Geodiversity in Pitsea holds the club's large Essex natural history collections which include a fine selection of Essex minerals, rocks and fossils. The centre is the culmination of an aspiration born in the early 1990s when the club had to step in after the closure of the Passmore Edwards Museum in Stratford and save the collections that its members had accumulated over the club's long history since 1880. A small number of specimens are currently on display but it is hoped that more will be accessible in the long term. The centre is currently open during weekends and certain other days with volunteer help – check online.

Chelmsford City Museum
Oaklands Park, Moulsham Street, Chelmsford, CM2 9AQ.
Telephone: 01245 605700.
The museum holds a large geological collection in storage. A few specimens are on show in a gallery with a tactile geological column wall display. Showcases plus video displays summarise a geological and archaeological history of Essex and the ice age. There is a large puddingstone, discovered locally, with an explanatory signboard outside the museum along the driveway.
A reference specimen cabinet has a useful display of rocks, pebbles and fossils found in Essex. It is there to help people wishing to identify their own specimens. For access to the cabinet, ask at the museum front desk. The cabinet was put together and donated by Essex Rock and Mineral Society; it includes specimens that were formerly on public display in the museum's geology room.

Colchester Natural History Museum
High Street, Colchester, CO1 1DN.
Telephone: 01206 282941
A range of fossils and other useful geological specimens is on public display in the gallery. The museum holds a large geological collection in storage.

Ipswich Museum
High St, Ipswich IP1 3QH. Telephone: 01473 433551
A gallery contains a display of Suffolk geology and some important specimens that demonstrate Suffolk and Essex geology; an online database provides many images of the museum's excellent storage collections of fossils and other geological materials, many of which are relevant to Essex geology. The museum galleries are currently (2022) scheduled for thematic redevelopment.

Saffron Walden Museum
Museum Street, Saffron Walden, CB10 1JL.
Telephone: 01799 510333
The museum holds a large geological collection and has a purpose-built geology gallery with good examples of fossils, explanations and displays of geological history, a replica of the Ashdon Meteorite and, outside the museum entrance, an enormous glacial erratic of a Jurassic septarian concretion.

Southend Central Museum
Victoria Avenue, Southend-on-Sea, SS2 6ES.
Telephone: 01702 434449
The thematic display includes a few geological and mineral specimens. A large collection of geological specimens is in storage; this includes the south Essex mineralogy collection of the former Passmore Edwards Museum. These can be visited and seen on request by researchers. There is no enquiry service.

Sedgwick Museum of Earth Sciences
Downing Street, Cambridge, CB2 3EQ.
Telephone: 01223 333456
One of the world's major collections of fossils, including many useful examples on display. Museum staff are willing to help with the identification of fossils, and to advise on their treatment and care.

Index

References to figures and photographs appear in *italic* type

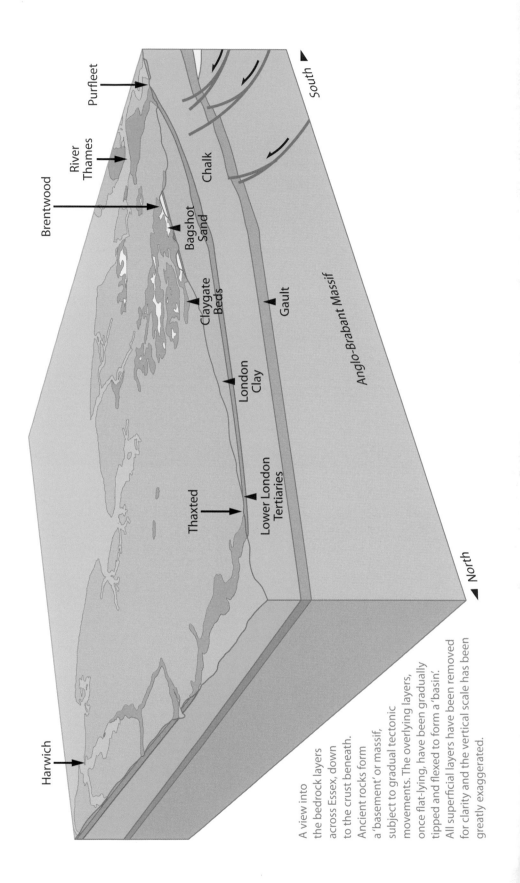

Purfleet

River
Thames

Brentwood

Chalk

Bagshot
Sand

Claygate
Beds

Gault

London
Clay

Anglo-Brabant Massif

Thaxted

Lower London
Tertiaries

Harwich

South

North

A view into
the bedrock layers
across Essex, down
to the crust beneath.
Ancient rocks form
a 'basement' or massif,
subject to gradual tectonic
movements. The overlying layers,
once flat-lying, have been gradually
tipped and flexed to form a 'basin'.
All superficial layers have been removed
for clarity and the vertical scale has been
greatly exaggerated.

Bedrock:

☐ Bagshot Sand
▨ Claygate Beds
▨ London Clay
☐ Lower London Tertiaries
☐ Chalk

Superficial layers:

☐ Alluvium
▨ Brickearth/coversand
▨ Post-Anglian river deposits
▨ Anglian: glacial outwash
☐ Anglian: glacial till
▨ Pre-Anglian river deposits
■ Crag group

Ipswich

Harwich

Walton-on-the-Naze

Clacton-on-Sea

Colchester

Sudbury

Braintree

Witham

Maldon

Chelmsford

Southend-on-Sea

Basildon

Saffron Walden

Bishop's Stortford

Brentwood

Grays

Harlow

Hornchurch

Dartford

Ilford

Finchley

Geological time chart

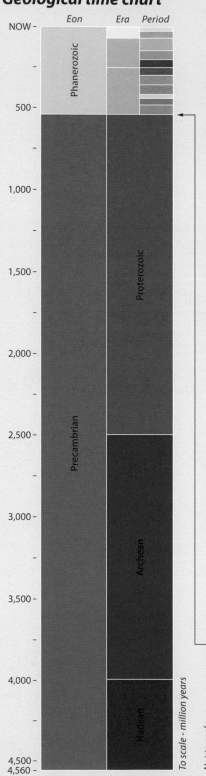

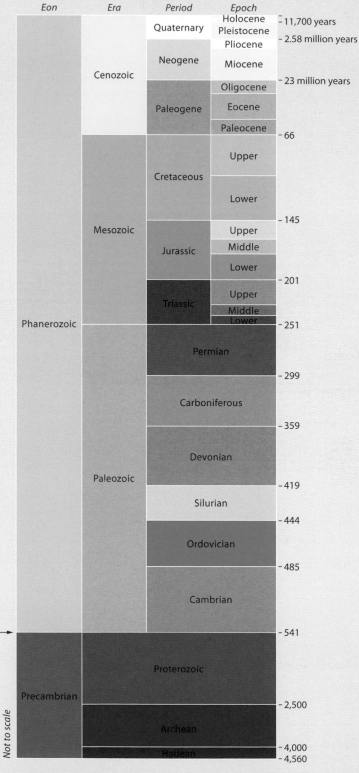

	Eon	Era	Period	Epoch	
			Quaternary	Holocene	11,700 years
				Pleistocene	
		Cenozoic	Neogene	Pliocene	2.58 million years
				Miocene	
			Paleogene	Oligocene	23 million years
				Eocene	
				Paleocene	66
	Phanerozoic	Mesozoic	Cretaceous	Upper	
				Lower	
					145
			Jurassic	Upper	
				Middle	
				Lower	201
			Triassic	Upper	
				Middle	
				Lower	251
		Paleozoic	Permian		
					299
			Carboniferous		
					359
			Devonian		
					419
			Silurian		
					444
			Ordovician		
					485
			Cambrian		
					541
	Precambrian		Proterozoic		
					2,500
			Archean		
					4,000
			Hadean		4,560

To scale - million years

Not to scale

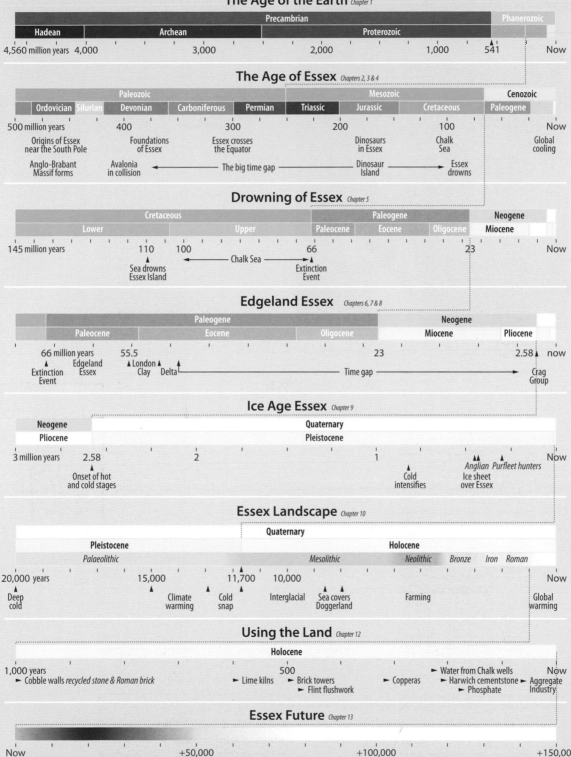

The Age of the Earth *Chapter 1*

| | | Precambrian | | | Phanerozoic |
| Hadean | Archean | | Proterozoic | | |

4,560 million years   4,000   3,000   2,000   1,000   541   Now

The Age of Essex *Chapters 2, 3 & 4*

| Paleozoic | | | | | | Mesozoic | | | Cenozoic |
| | Ordovician | Silurian | Devonian | Carboniferous | Permian | Triassic | Jurassic | Cretaceous | Paleogene |

500 million years   400   300   200   100   Now

Origins of Essex near the South Pole — Foundations of Essex — Essex crosses the Equator — Dinosaurs in Essex — Chalk Sea — Global cooling

Anglo-Brabant Massif forms — Avalonia in collision ← The big time gap → Dinosaur Island — Essex drowns

Drowning of Essex *Chapter 5*

| Cretaceous | | | Paleogene | | | Neogene |
| Lower | | Upper | Paleocene | Eocene | Oligocene | Miocene |

145 million years   110   100   66   23   Now

Sea drowns Essex Island ← Chalk Sea → Extinction Event

Edgeland Essex *Chapters 6, 7 & 8*

| Paleogene | | | Neogene | |
| Paleocene | Eocene | Oligocene | Miocene | Pliocene |

66 million years   55.5   23   2.58   now

Extinction Event — Edgeland Essex — London Clay — Delta ← Time gap → Crag Group

Ice Age Essex *Chapter 9*

| Neogene | Quaternary |
| Pliocene | Pleistocene |

3 million years   2.58   2   1   Now

Onset of hot and cold stages — Cold intensifies — Anglian Ice sheet over Essex — Purfleet hunters

Essex Landscape *Chapter 10*

	Quaternary	
Pleistocene	Holocene	
Palaeolithic	*Mesolithic*	*Neolithic* *Bronze* *Iron* *Roman*

20,000 years   15,000   11,700   10,000   Now

Deep cold — Climate warming — Cold snap — Interglacial — Sea covers Doggerland — Farming — Global warming

Using the Land *Chapter 12*

| Holocene |

1,000 years   500   Now

► Cobble walls *recycled stone & Roman brick*
► Lime kilns   ► Brick towers   ► Copperas   ► Water from Chalk wells   ► Aggregate Industry
► Flint flushwork   ► Harwich cementstone
► Phosphate

Essex Future *Chapter 13*

Now   +50,000   +100,000   +150,000 years

► Anthropocene heating event and extinctions
► Floods over Essex lowlands
Onset of next cold stage — Next glaciation   Essex joined to Europe by Doggerland — Essex steadily moving northward

Ian Mercer worked for the British Geological Survey and Geological Museum for 25 years and was Director of Education for the Gemmological Association of Great Britain for nearly two decades. **Ros Mercer** was a geology and physics teacher for more than 20 years. They have been leaders for the Essex Rock and Mineral Society for many years; UK national advisors for U3A Geology; and officers for GeoEssex – the county geoconservation steering group. In 2020, they were awarded together the Halstead Medal by the Geologists' Association for their achievements as 'ambassadors in the promotion of geology, contributing to education at all levels'. Ian Mercer is the author of several books on related subjects.